The Process of Innovation

a Study of Companies in Canada, the United States and the United Kingdom

by

Nuala Swords-Isherwood

Principal Research Fellow

The Technical Change Centre, London

PART I: A REPORT ON THE PROCESS OF INNOVATION IN INDUSTRY

PART II: INDUSTRY CASE STUDIES

BRITISH-NORTH AMERICAN COMMITTEE

Sponsored by

British-North American Research Association (UK)

National Planning Association (USA)

C. D. Howe Institute (Canada)

*British-North American Committee

ISBN 0-902594-43-5

Published by the British-North American Committee

Printed and bound in the United Kingdom by
Contemprint Limited, London SE1

October 1984

This report has been prepared in conjunction with The Technical Change Centre, London

Contents

Statement of the British-North American Committee to Accompany the Report

Discussions at recent meetings of the British-North American Committee have convinced us that unless the older industrial countries, including the U.S., U.K. and Canada, can increase their success rate in translating industrial research into commercially viable products or process innovations, they will in many sectors, become progressively less competitive in international markets. Although this can be evaded for some time by various protectionist measures, it will eventually lead to lower levels of economic activity, fewer job opportunities and a generally lower standard of living.

A study of industrial innovation in our three countries by Dr. Kerry Schott*, which we published in July 1981, pointed out that the relatively poor economic performance by the U.K., U.S. and Canada in the 1970s had discouraged industrial innovation which was, in turn, leading to an erosion of competitiveness. Schott shows that the proportion of resources devoted to research and development in the three countries was declining and compared unfavourably with our principal competitors. She warned of the consequences if the trend was not reversed.

With the objective of understanding more clearly what are the significant factors which lead to successful innovations by companies, the British-North American Committee asked The Technical Change Centre in London to undertake an inter-country/inter-sector comparison on the process of innovation. For this Dr. Nuala Swords-Isherwood, a principal research fellow at the Centre, selected thirty-seven large and small companies in the three countries for interview and analysis. This is her Report which we warmly commend for wider study. It does not tell us how to be successful, but it does show that the important elements of success are remarkably similar both between countries or even economic sectors.

We endorse her main conclusion that the role of management in successful innovation is crucial. For a company to have a research function is not a sufficient condition for success. Only when top management delegates equal "effective power" to its managers involved in the innovation process, and when adequate and continuing resources are allocated to design and development, is a firm likely to be a consistent and successful innovator.

In this respect the selection of projects, their cost control, cut off points, key objective and progress indicators seem to us to be among the essential management tools required for success, regardless of the country or sector in which the company operates. Companies must accept that product and process development involves a substantial element of risk; it can be reduced to some extent by thorough assessment of market needs but can never be eliminated.

We also agree with the author that it is unrealistic to try to emulate a model of management which is foreign to existing forms of organisation just because it appeared to be successful within its own culture in producing impressive innovations.

However, we do call on our respective Governments to look again at the legal and fiscal conditions which seem to impede the chances of innovations being brought to fruition on a timescale at least as advantageous as that of our main competitors.

If more industries and companies in the U.S., U.K. and Canada were to take the issue of sustained product and process innovation seriously and structure themselves in ways which ensure that potential "winners" are not stifled soon after birth, we are confident that we shall be able to reverse the drift towards industrial decline – with all that this implies for the future economic and social life of our nations.[1]

---

* Kerry Schott, *Industrial Innovation in the United Kingdom, Canada and the United States,* BNAC, London, Toronto, Washington, 1981.

1. *I do not think it at all likely that reversing the drift toward industrial decline can be achieved by re-structuring to avoid stifling winners. This is but one of a range of changes which in my opinion, would be necessary – some of which are within the power of both industries or companies to accomplish –* ***Ingram Lenton***

Footnotes to the Statement as a whole

Firstly, I regret that the term "R & D" is extensively used for this gives the impression that 'Research and Development" is a single activity whereas in fact over a wide area of industry they are distinct and very different activities in objective, cost and outlook required for success. The objective of Research is increase of knowledge which in industry will usually be relevant to a particular part of the business, whereas Development is closely related to Design and is the process of proving a design, whether of a product, process or system so that it meets in all respects the market specification, including cost and time scale, and will be saleable at a profit. Well conducted Research promotes efficient Design and Development but the cost of the latter is an order of magnitude greater than that of Research and requires an entirely different outlook, closely related to market forces.

Secondly, there is insufficient emphasis in the report on the importance of Project Management in successful innovation. The difficult stage is the conversion of an idea, through design and development, into something that is marketable and profitable. This requires high quality of judgement and the authority to apply it effectively, through quick decision making, to the many problems which arise in the Design and Development stages. Sometimes they may be so great, time consuming or expensive to solve that the project must be abandoned: but if the project goes ahead, and is to be profitable, Development, leading to production and marketing, must proceed as rapidly and smoothly as possible. This will only be achieved if a Project Director, Manager or "Champion" is appointed with clear responsibility coupled with full authority within an agreed budget, to push the project through to a successful conclusion.—**Lord Caldecote**

The differences of opinion on innovative strategy within an organization are not only the differences among the functional groups of research, engineering, marketing and finance. There may also be major differences between groups within the R & D structure in which the senior members represent an older technology in which they were trained and the younger members represent a newer technology in which they have been trained during their university years. For example, classical optics and photography is an older technology while solid-state electro-optics is a new technology possessing a much greater potential for the future in fields important to the company. In order that the organization as a whole becomes committed to the newer technology it may become necessary to place younger researchers and engineers in key positions in order that the newer technology is accepted by the firm as a whole.—**Franklin Lindsay**

A very considerable amount of corporate restructuring, cost-cutting and re-organization of both a sectoral and a corporate nature has been taking place in the United States in the past three years or so, with increased competitive capability flowing therefrom. This increased competitive power has not yet manifested itself in export performance; on the contrary, it has been obscured by the high value of the U.S dollar in relation to other major currencies. The consequence is colossal deficits on trade account rather than surpluses. If and when a significant realignment of exchange rates takes place, it may produce some increasingly difficult competitive problems for Europe and possibly even for Japan.

This is to say, in short, that while industrial research and product innovation are obviously of key importance to the fundamental competitive quality of our national economies, other factors of a macro nature can be dominant over periods of time. In organizing our world economy, we may not easily gloss over the fact that exchange rates, for example, can be more central than research and innovation.—**Nathaniel Samuels**

Members of the Committee Signing the Statement

J. A. ARMSTRONG
Toronto, Ontario

DAVID ATTERTON
Chairman, Foseco Minsep Ltd.

CHARLES F. BAIRD
Chairman and Chief Executive Officer, INCO Limited

ROBERT A. BANDEEN
Chairman and President, Crown Life Insurance Company

JOSEPH E. BAIRD
Los Angeles, California

PROFESSOR SIR JAMES BALL
Chairman, Legal & General Assurance Group plc.

SIR DONALD BARRON
Chairman, Midland Bank plc.

MICHAEL BELANGER
Chairman of the Board, President and Chief Executive Officer, National Bank of Canada

W. J. BENSON
Deputy Chairman, National Westminster Bank plc.

ROGER BEXON
Deputy Chairman, The British Petroleum Company plc

CARROL D. BOLEN
Vice-President, Pioneer Hi-Bred International Inc.

*VISCOUNT CALDECOTE
Chairman, Investors in Industry Group plc.

SILAS S. CATHCART
Chairman, Illinois Tool Works Inc.

GEORGE J. CLARK
Executive Vice President, Citibank N.A.

DONALD M. COX
Director and Senior Vice President, Exxon Corporation

JAMES W. DAVANT
Retired Chairman, Paine Webber Incorporated

RALPH P. DAVIDSON
Chairman, Time Incorporated

WILLIAM DODGE
Ottawa, Ontario

WILLIAM H. DONALDSON
Chairman and Chief Executive, Donaldson Enterprises Inc.

EDWARD DONLEY
Chairman, Air Products and Chemicals Inc.

JOHN DU CANE
Director, AMAX Inc.

KEN DURHAM
Chairman, Unilever plc.

GERRY EASTWOOD
General Secretary, Association of Patternmakers and Allied Craftsmen

HARRY E. EKBLOM
Chairman, University Science Partners Inc.

GLENN FLATEN
President, Canadian Federation of Agriculture

C. S. FLENNIKEN
Chairman, President and Chief Executive Officer, CIP Inc

Sir ALISTAIR FRAME
Deputy Chairman and Chief Executive, Rio-Tinto Zinc Corporation

SIR CAMPBELL FRASER
President, Dunlop Holdings plc.

ROBERT R. FREDERICK
President, RCA Corporation

NALCOLM GLENN
Executive Vice President, Reed Holdings Incorporated

PETER A. GRAHAM
Senior Deputy Chairman, Standard Chartered Bank plc.

JOHN H. HALE
Managing Director, S. Pearson & Son plc.

HON. HENRY HANKEY
British Secretary, BNAC, Westerham, Kent

FRED L. HARTLEY
Chairman and President, Union Oil Company of California

PAUL HAZEN
Vice Chairman, Wells Fargo & Company

HOWARD D. SAMUEL
President, Industrial Union Department, AFL-CIO

*NATHANIEL SAMUELS
Advisory Director, Lehman Brothers Kuhn Loeb Inc.

LORD SEEBOHM
Dedham, Essex

THE EARL OF SELKIRK
President, Royal Central Asian Society

JACK SHEINKMAN
Secretary-Treasurer, Amalgamated Clothing & Textile Workers' Union

LORD SHERFIELD
Chancellor of Reading University

R. MICHAEL SHIELDS
Managing Director, Associated Newspapers Group plc.

GEORGE L. SHINN
Chairman and Chief Executive Officer Emeritus, The First Boston Corporation

GORDON R. SIMPSON
Chairman, General Accident Fire and Life Assurance Corporation plc

SIR ROY SISSON
Chairman, Smiths Industries plc.

SIR LESLIE SMITH
Chairman, BOC International plc

RALPH I. STRAUS
New York, NY

THOMAS H. B. SYMONS
Vanier Professor, Trent University, and Director, Celanese Canada Inc.

JAMES C. THACKRAY
Chairman, Bell Canada

W. C. THOMSON
A Managing Director, Royal Dutch/Shell Group of Companies

ALEXANDER C. TOMLINSON
President, National Planning Association

ALAN TUFFIN
General Secretary, Union of Communication Workers

WILLIAM I. M. TURNER, JR.
Chairman and Chief Executive Officer, Consolidated-Bathurst Inc.

W. O. TWAITS
Toronto, Ontario

JOHN A. URQUHART
Executive Vice President and Sector Executive, International Sector, General Electric Company

MARTHA REDFIELD WALLACE
President, Redfield Associates

GLENN E. WATTS
President, Communications Workers of America, AFL-CIO

WILLIAM L. WEARLY
Chairman, (Retired), Ingersoll-Rand Company

VISCOUNT WEIR
Chairman, The Weir Group Limited

CHARLES WOOTTON
Director, International Public Affairs, Gulf Oil Corporation

*See footnotes to statement

Author's Acknowledgements

I am grateful to the British-North American Committee (BNAC) and BNAC members for supporting this work and so generously giving their time and effort in the course of it. The work would not have been possible without the cooperation of the many industrialists involved. I received many helpful comments on the drafts of the report but would like to thank Simon Webley and Sperry Lea in particular for their thoughtful suggestions. I would like to thank Jenny Bryan-Brown, Maryon Jacobs and most particularly Katherine Woolley for all their work. A special word of thanks must go to Baron Isherwood without whom my international travels would be impossible. Of course the responsibility for the contents of the paper is all my own.

NUALA SWORDS-ISHERWOOD

The Technical Change Centre, London

The Process of Innovation: A Study of Companies in Canada, the United Kingdom and the United States

Nuala Swords-Isherwood

SUMMARY

Concern with industrial innovation and its role in competitiveness is now widespread. This study, reflecting this concern, is of the process of innovation in the UK and North America. It provides a micro-economic picture of innovative activity in a selection of companies in different industrial sectors in three countries, and follows up the 1981 British-North American Committee work on innovation by Kerry Schott.

Twenty-one companies contributed to the case studies on innovation included in Part II, and a further 16 to the discussions on R & D and innovation. The original participants were members of the British-North American Committee. The study was conducted by means of interviews. A union view was also obtained.

The report analyses some of the literature on innovation and illustrates the wide agreement on factors likely to assist successful innovation and on those likely to impede it. The study suggests the importance of innovation in changing an industry's basis of competition, and in disturbing established systems within it. The crucial role of finance in the process of innovation is discussed.

The report explores the relationship between R & D and innovation both as it is explained in the literature and from interviews with company officials in the electrical/electronic and pharmaceutical industries. In these industrial sectors the importance of research and development, and recognition of the importance of R & D in innovation, is total. However, even in these sectors, good R & D, although necessary, is not sufficient for innovative success.

In the five companies in the insurance industry study, the essential role of data processing and its contribution to innovation was heavily stressed. Reliance on a single main equipment supplier was marked. Despite this, differences of emphasis, and varied legislative environments, technical advances in data processing were now used not just to increase efficiency and reduce costs, but also for product innovations. Rapidly increasing competition in the industry was accelerating the rewards of successful innovation and the costs of not innovating.

Within the paper industry, although there were opportunities for both product and process innovation, the pressures under which these opportunities must be taken were considerable. The cyclical nature of the industry, its maturity, its vulnerability to variable financial factors, and a highly competitive business environment, make innovation more necessary but more problematic. The primary danger was one of over emphasis on cost which could have serious implications for achieving the appropriate balance of management skills necessary for successful innovation.

The problems of the paper industry to an extent faced all the materials and resources suppliers in the study, but the severity and immediacy of the danger related to the financial strength of the sector. Thus the innovative company in the mining and metals sector had had to retrench and restructure its efforts while attempting to maintain its innovative thrust. The other company in this pair felt reassured in its strategy of seeking to be efficient and competitive rather than innovatory. The oil companies in the study had also reorganized their research and innovation efforts. Interestingly the UK company had decided to move into more speculative areas, but with business objectives always in mind, while one US company had decided to concentrate its efforts on its primary market and to seek innovations in more limited fields.

The machine tool companies in the study, although faced with rapid and dramatic technical and competitive challenges, were attempting to face these new pressures with increased investment in R & D, and a new awareness of the importance of innovation. Hit both by the recession and the threat from relatively new competitors, seizing technological opportunities and increasing market sensitivity were seen as fundamental to success in innovation.

In the sunrise industries the pace of change was rapid. Biotechnology in the UK has been dominated by large, diverse firms, well established in other fields, while in the US large numbers of small companies have developed, many surviving on venture capital. In both countries the general economic climate was seen to have adversely affected innovation in the industry as had the lack of thorough commitment.

The medical imaging companies had grown rapidly. The UK company initially had some problems in matching its business and marketing capacity to its technical capability. Both companies now recognised the dangers of over emphasis on technical innovation and achievement at the expense of commercial viability, and this involved increased awareness of market requirements.

The microelectronics companies, located in a highly innovative, rapidly changing, competitive environment, felt under continuous pressure to innovate. For them, small company size was important in fostering and supporting creative people, although acquiring people of the right calibre caused problems.

Overall international differences were of less significance than might have been expected from the literature and sectoral differences emerged as of immense importance. The nature of the sample is probably of relevance here, but in this study the opportunities for innovation and the constraints upon it were often similar in different countries, but different in different industrial sectors.

The Report reinforces the crucial role of management in successful innovation and suggests the necessity for a balance of management functions, both in terms of visibility and power. Achieving an appropriate mix of management functions is far from easy. Once this balance has been attained, innovation, the report suggests, could then be fostered through positive creative conflict amongst existing functional divisions. This would be preferable to expecting orientation and direction to emerge through consensus in a quasi-Japanese fashion.

The report also suggests that the appropriate functional balance of management is related to the level of technology, but is primarily derived from the industrial sector in which the company is located. Further studies of the process of innovation within sectors could allow the development of a checklist of imbalances in each sector, and their causes. This study then, might be seen as a step towards development of an analytic framework which could be used to facilitate successful innovation.

PART ONE:

A REPORT ON THE PROCESS OF INNOVATION IN INDUSTRY

I. Introduction and Background to the Study

Concern with industrial innovation and its role in competitiveness is now widespread. Many economic studies indicate that the contribution of R & D and innovation to economic growth and productivity is high. Those who suggest that a lack of R & D and innovation is not a cause, but a consequence of the current economic malaise, agree that stimulating innovation will make escape from these adverse circumstances easier. Hence industrial innovation, although not the sole determinant of economic growth, makes an important contribution to it[1].

This study is concerned with the process of innovation in the UK and North America; its aim is to provide a microeconomic picture of innovative activity by looking at a selection of companies in different industrial sectors in the US, UK and Canada. It follows the 1981 British-North American Committee study on innovation, by Kerry Schott, published in 1981[2].

Schott emphasised the inter-relationship between the state of the economy and the level of innovation activity. She pointed to the relatively poor economic performance of the UK, Canada and the USA and claimed that it discouraged industrial innovation, and that relative decline in these three countries contributed to further economic decline. She showed that the comparatively slow productivity growth rates in Britain and North America were declining faster than those of their competitors. Investment per capita was also shown to be relatively low in these countries, and unemployment rates relatively high. This, it was suggested, was particularly the case in the UK. Indicators of innovation suggested that the share of resources devoted to it in the three countries in the

---

1 For an interesting discussion see Schacht, W. H. *Industrial Innovation and High Technology Development,* Issue Brief Number IB0005, The Library of Congress Congressional Research Service Major Issues System, Washington D.C., 1983.

2 Schott, K. *Industrial Innovation in the United Kingdom, Canada and the United States,* British-North American Committee, London, 1981.

study had declined and compared unfavourably with those in France, West Germany and Japan. Schott talked of the "circularity of influence", that is, the general economic climate that affects innovation which itself influences economic performance. She suggested that on the basis of the evidence examined, it was imperative that innovation be encouraged in the UK, Canada and the USA.

For Schott, a policy of encouraging innovation was vital if industries in these countries were to recover from their depressed levels and compete effectively in world markets. But policies to stimulate innovation, she argued, must be part of a policy package to stimulate economic recovery. The British-North American Committee felt that this macro-economic picture, with its policy suggestions, could usefully be followed by a microeconomic analysis of the innovation process. The expectation was that one might see how innovation really works, how money is being spent within companies, how indeed it is possible to organise for change.

AIMS AND METHODS OF THE STUDY

This study examines the process of innovation through a broad range of innovations, often minor, which were felt to be commercially significant by the companies visited. These were identified as the most interesting innovations in recent years by company respondents. Some of these innovations were the first uses of a technical change and new to the economy, but many were new only to the firm and were, therefore, instances of innovation diffusion. No attempt has been made to make distinctions among the innovations as to their significance. The firm's evaluation of their importance has been accepted. The necessary context to allow for some judgements has been provided by the pairing and grouping of companies. The comparisons are less than perfect because of the size of the companies involved and the number of their interests. The emphasis on the behaviour of the firm recognises their central role in the innovation process[3].

---

3 See: W. J. Abernathy and P. Townsend, "Technology Productivity and Process Change", *Technology Forecasting and Social Change,* (7) 1975; and Nelson, R. R. and Winter, S. G. *An Evolutionary Theory of Economic Change,* Harvard University Press, Cambridge, M.A., 1982.

During the course of the project, it was agreed to add an additional element concerning the organisation of research and development in some major companies in the three countries. This was an attempt to clarify the relationship between R & D and innovation which, although obviously intimate, is often obscure. Clearly one innovation, even one successful innovation, does not necessarily make a company innovative. It was felt therefore that a study of the organisation of R & D in large companies, which must maintain a consistent and constant research effort over a long period of time to maintain their competitive positions, would help to further our understanding of what is involved in innovation.

The participants in the study initially were British-North American Committee members who then suggested additional comparable companies for their groupings, which were then contacted direct. The companies involved in the sector analysis of R & D and innovation had participated in an earlier study of R & D and recession in the UK, and Canadian and US comparisons were sought both from amongst committee members and directly[4]. Twenty-one companies contributed to the case studies and a further 16 to the discussions on R & D and innovation. A full list of participating companies is given in Appendix 1. The industry sectors involved were: pharmaceuticals, electronics, insurance, paper, mining and metals, oil, machine tools, biotechnology and medical imaging. The cases have been grouped to illustrate innovation in a service industry, in the materials and resources area, in a traditional manufacturing industry, and in the so called 'Sunrise' industries. This mix allows comparison of the innovation process in different industrial sectors with different markets and different degrees of technological sophistication[5].

The study was conducted by interviewing one or more respondents in each company, usually the managing director and, or, the research director who had a thorough knowledge of the process of innovation with the firm. (For interview outline see Appendix II).

Although innovation and innovation diffusion have been subject to much study both in North America and the UK, studies involving international comparisons are less common. Empirical studies tend to

---

4 Swords-Isherwood, N. *Research and Development in Recession in British Industry*, Technical Change Centre, Mimeo, London, 1983.

5 For a discussion of the importance of sectoral differences see R. R. Nelson, and S. G. Winter, "In Search of Useful Theory of Innovation", *Research Policy*, (6) 1977.

be either of major innovations or individual cases or paired cases of success and failure. This study provides a somewhat different perspective since it is international, concerned with the process of innovation within companies, and cases are grouped to allow a better picture of innovation activity in particular industrial sectors to emerge. Many of the conclusions drawn will of course reinforce the findings of other studies, but the suggestion of developing sectorally specific formulae for avoiding innovative failure, although not a formula for success, is perhaps more specific than is usual. The next sections of the report contain brief analysis of some of the literature on innovation, and explore the relationship between innovation and R & D. The conclusions of the case studies are then presented, and the overall conclusions end Part I of the report. The case studies are presented in Part II.

II. The Nature of Industrial Innovation

Invention is focussed on the creative and absolutely new idea, product, process, or technology, which is not necessarily of commercial use. Innovation is a process which consists of the technical, industrial and commercial steps which lead to the marketing of new manufactured products or new services and the commercial use of new technical processes, products and services. Product developments represent additions or modifications to an existing product. Even though these three activities are notionally distinguishable, in practice it could be mistaken to over-emphasise any one, since they are intimately linked in successful competitive activity.

There are many studies of innovation both theoretical and empirical. Early models of innovation emphasised scientific and technological advance, and suggested that discoveries in basic science led to industrial technological developments. These technology push models would suggest that a customer might not know whether he had a need for the new discovery or invention until after it had been presented to him. Later models, often based on empirical studies, emphasised the importance of felt need or need pull, in which innovation followed from a perceived, and often market related need. There are weaknesses in these models. The technology push model suggested that more R & D expenditure resulted in more innovation, while the market pull models ignored the fact that market needs often result only in technological incrementalism.

The recognition of these weaknesses led to the development of interactive models, in which the process of innovation represents the confluence of technological capabilities and market needs within the framework of the innovating firms. Interactive models show innovation as a logically sequential, though not necessarily continuous, process, requiring a complex net of communications both intra and extra organisational, linking in-house functions within the firm and the firm to the scientific and technological community and the work of the market-place.

Within the framework, key characteristics of successful and unsuccessful innovators have been suggested by empirical work on innovation. Different studies place weight on different factors but when a number of studies are considered together, there seems to be strong evidence to suggest that not only are these individual factors of importance in innovations, but combinations of particular factors are also necessary for success.

One of the early studies of innovation by Carter and Williams[6] was an extensive study for the Board of Trade of innovation decisions of British firms. The first systematic data on the factors involved in the initiation of commercial innovation were presented. The stages in the generation and application of scientific knowledge from basic research to the point of substantial investment in the commercial production of an innovation were outlined, and the managerial and economic factors which facilitated the application of such knowledge on the part of private firms were also discussed. Their later work concerned the specific factors influencing the decisions of firms to invest in new technologies[7]. The authors concluded that explicit consideration of market demand factors is rare in decisions concerning the level of investment in research and development, because of an absence of clearly defined corporate objectives, and a focus upon short-term development and survival.

The authors examined in detail over 200 cases of product or process developments involving research and development departments. In 60 of these cases it was possible to state the source of the original impulse to undertake the R & D. Twenty-five per cent of the developments originated in a research and development department. Eighteen per cent arose from commercial foresight: that is, a forecast that due to market developments, the need for a new product or process would arise, and research staff then would be involved in developing it. Ten per cent arose from shorter term needs of the sales department. In 18% of the cases the origin was an overseas example or contract; in 14% the origin was consumer pressure (most frequently from a government agency); and 10% came from production departments.

Carter and Williams examined firms' decisions to proceed with development in 250 cases. They stated the following:

> In 65 cases . . . the firms were induced to proceed by the conviction that the potential market was large. In 54 cases improvement in quality of product was thought to improve

6 See Carter, C. F. and Williams, B. R. *Industry and Technical Progress: Factors Governing the Speed of Application of Science to Industry,* Oxford University Press, London, 1957.

7 Carter, C. F. and Williams, B. R. *Investment in Innovation,* Oxford University Press, London, 1959.

> the competitive position of the firm. In 69 cases the expectation of a saving in process cost was a sufficient motive. In 35 cases the fact that there was a 'significant innovation' was thought to be in itself a sufficient reason for going ahead. In 18 cases the innovation, by broadening the basis of the firms operations, was thought to improve the prospects of long-term stability. In 16 cases innovations were intended to guarantee supplies of materials. In 10 cases 'government interest' made the project seem worthwhile. In 13 cases the fact that the innovation 'increased output', was a sufficient incentive[8].

In 'Investment and Innovation' the authors distinguish between active and passive innovation decisions: "In making a passive innovation the firm is responding to direct market pressure, as shown by excess demand or by developing competition and falling profit margins . . . In making an active innovation, the firm is deliberately searching for new markets and techniques, even though there may be no direct market pressure to do so"[9]. Carter and Williams found that 60% of the 204 innovations considered were of the active variety.

Myers and Marquis in an often quoted study of successful innovations suggested that innovation is not a single action but a total process[10]. Technical change was to a significant extent seen as based on the cumulative effect of small, incremental innovations. In their study innovations arising from primary market factors were relatively more frequent in small companies, and innovations arising from technical or production factors were more frequent in larger companies. They suggested that market related innovations were more likely to be product rather than process innovations and less likely to require inventive adaptations or an original new production process. Recognition of demand was seen as a more frequent factor in innovation than recognition of technical potential.

Myers and Marquis concluded that minor improvements could be programmed but major innovations were quite unpredictable, high risk ventures, sometimes appearing accidently, and outside the industrial laboratory. Innovations could be offensive or defensive.

---

8 Carter and Williams, 1957, Op.cit.

9 Carter and Williams, 1959, Op. cit.

10 Myers, S. and Marquis, D. G. *Successful Industrial Innovations,* National Science Foundation, Washington D.C., 1964.

The process of innovation must be seen as managed and it required technical competence, experience, and effective communication for success. "The management of innovation is a corporate-wide task, and is too important to be left to any one specialised functional department[11]."

Langrish, Gibbons, Evans and Jevons investigating 84 UK innovations which were granted the Queen's award for technological innovation in 1966 and 1967, identified seven factors of importance to the firms success[12].

1. A top person—the presence of an outstanding person in a position of authority
2. Other person—other type of outstanding individual
3. Clear identification of a need
4. The realisation of the potential usefulness of the discovery
5. Good co-operation
6. Availability of resources
7. Help from government sources

The following six factors caused delay to an innovation.

1. Some other technology not sufficiently developed
2. No market or need
3. Potential not recognised by management
4. Resistance to new ideas
5. Shortages of resources
6. Poor co-operation or communication

The Sappho study (Scientific Activity Predictor from Patterns with Heuristic Origins) examined 29 pairs of similar projects, one of which was successful, the other less so, and it suggested the following characteristics[13]:

1. Successful innovators were seen to have a much better understanding of user needs.
2. Successful innovators pay more attention to marketing.

---

11 Myers and Marquis, 1964, Op.cit.

12 Langrish, J. et al. *Wealth from Knowledge, A Study of Innovation in Industry*, Macmillan, London, 1972.

13 *Success and Failure in Industrial Innovation*, Project Sappho, Science Policy Research Unit, University of Sussex, 1972.

3. Successful innovators perform the development work more efficiently than failures though not necessarily more quickly.
4. Successful innovators make more effective use of outside technology and advice, even though they perform more of the work in-house.
5. Responsible individuals in the successful attempts are usually more senior and have greater authority than their counterparts who fail.

A 1971 study of 53 projects which had been abandoned during development suggested that the causes of abondonment be divided into two categories, environmental factors and organisational factors[14]. Under environmental factors the following were listed: unattractively small markets, uncertainty with monopsonistic buyers, an unattractive level of competition, uncertainty of suppliers, and obsolescence. Under organisational factors came these: lack of marketing capacity or expertise, lack of production capacity or expertise, faulty communications with associated firms, R & D cost escalation, and shortage of R & D resources.

Rothwell has compiled a list of the characteristics of successful innovators and technically progressive firms based on the main results of nine major empirical studies[15]. Together they represent the study of over 1,000 industrial innovations. Successful innovators, he suggests, recognise the importance of the six factors listed here.

1. Commitment to innovation
2. Innovation as a corporate wide task
3. Attention to marketing, user needs and after sales servicing
4. Efficient design and development work
5. Good internal and external communication
6. Management skills and professionalism

Probably the product champion who has both technical knowledge and enterpreneurial flair should be added to the list.

14 *On the Shelf: A Survey of Industrial R&D Projects Abandoned for Non-Technical Reasons,* Centre for Industrial Innovation, London, 1971.

15 R. F. Rothwell, "Policies anrl Industry", in *Technical Innovation and British Economic Performance,* Ed. K. Pavitt, Macmillan, London, 1980.

A recent interesting theoretical contribution to the innovation debate develops the idea of "transilience[16]." An industry's structure, basis of competition, and nature of resources are all influenced to some degree by technical change. The degree of change depends on the nature and extent of an innovation's "transilience": that is, its capacity to influence production systems, and its linkages with the market. Figure 1 shows the evolution of technology transilience.

Figure 1

EVOLUTION OF TECHNOLOGY TRANSILIENCE

Impact on Market Linkages

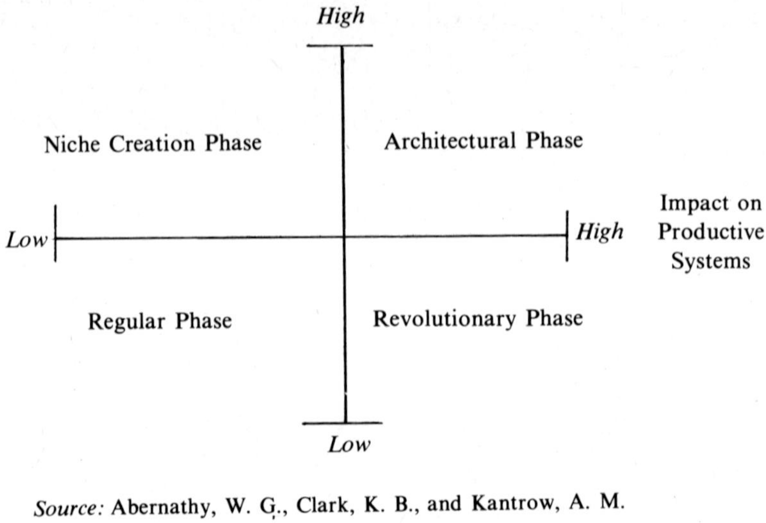

Source: Abernathy, W. G., Clark, K. B., and Kantrow, A. M. (see footnote 16).

An architectural innovation creates entirely new design concepts and the production systems and markets to accommodate them. The more architectural innovations require substantial change in existing organisational commitments, the more obvious their transilience. In the regular phase of innovation, technical changes refine and extend established designs in products and processes. The cumulative impact

16 Abernathy, W. G., Clark, K. B., and Kantrow, A. M. *Industrial Renaissance—Producing a Competitive Future for America,* Basic Book, New York, 1982.

of these incremental innovations may be very substantial. The regular phase of innovation does not affect extant market linkages, but the niche creation phase does. The latter involves the re-packaging of established technology in support of efforts to reach new markets. Eventually regular and niche creating innovations no longer give rise to competitive vitality and new technology then must be applied to existing markets in the "revolutionary" phase of innovation.

The case studies that form the basis of the BNAC report provide examples across the range of this matrix. Abernathy, Clark and Kantrow maintain that "What is truly important about innovation, whatever its technical novelty, is the extent to which it changes an industry's basis of competition at the same time that it disrupts established production competence, marketing and distribution systems, capital equipment, organizational structures, and the skills of both managers and workers[17]".

17 Abernathy, Clark and Kantrow, 1982, Op.cit.

III. Finance and Innovation

Innovation, businessmen assert, is about using creative ideas to make money. Translating such ideas into reality may cost a great deal. Clearly there are financial constraints and problems which may set limits on innovation. Equally, for an innovation to be successful, the acquisition of adequate financing to cover the research and development is as vital as the original idea.

Intuitively the notion that problems of finance are likely to slow down innovation is appealing, as Carter and Williams suggest: "The idea that action seen to be good is prevented by shortage of money comes very easily to the mind, for it fits in with one's private belief that one could make use of a larger pay cheque[18]."

The cost of research, development and innovation can be enormous, more than $80 millions (US) to bring a new drug to market, in the region of $2,000 millions to produce a new car, and $4,000 millions or $5,000 millions to produce a new aeroplane. The heaviest burden of these enormous costs will fall at the development end of the cycle. Many would suggest that such innovations are too expensive for an industrial company to contemplate and that innovation would be furthered better by cooperative arrangements: sharing knowledge becomes sound business sense. Innovations can also cost very little, or the cost can be insignificant to a major company but formidable to a small one.

The problems of the smaller company in getting finance have been analysed many times. Financial institutions, particularly in the UK, are often heavily criticised for their unwillingness to invest in small high technology firms. Usually this is said to be a reflection of the institutions' lack of science and technology background. Possibly, on occasion, this unwillingness has a sound basis and these difficulties have been reported even in the earliest studies. Carter and Williams reported, "It was suggested to us that the common reason for the failure of a small borrower to be credit worthy is that he does not command the diversity of qualities needed in successful business. He is perhaps a brilliant technician, but no manager of men; strong on production, but weak on marketing; or knowledgeable of his trade, but innocent of the mysteries of finance[19]".

---

18 Carter and Williams, 1957, Op.cit.

19 Carter and Williams, 1957, Op.cit.

One development, partly in response to such problems, has been the growth of venture capital markets. Venture capital has most obviously flourished in the US where structural factors such as a suitable tax system, mobility of highly qualified people, customers receptive to new ideas and products, and also cultural factors, such as respect for individualism and profit making, have provided a benign environment. There is also venture capital activity in the UK and it is suggested that venture firms of an American type will achieve greater significance in the British economy. The people directing such firms are likely to continue to come from outside the city, that is from outside mainstream financial institutions.

It is easy to exaggerate the role of venture capital. Even in the US venture capital firms are "... small organizations with no more than a dozen people working for them. It follows from this that the entire venture capital community, as it is so often styled by those that belong to it, consists of only a few thousand. Within the larger circle the mere three or four hundred people associated with the leading firms constitute the centre of real influence[20]".

The contribution of venture capital firms to the creation of new industries in the United States has been considerable, both in mobilisation of resources and in the stimulation of managements, but critics suggest that such innovations are in a limited field and that the effect on the US economy has been marginal. Perhaps more importantly, it has been suggested that the efforts of venture capital firms may weaken older larger firms, in particular by encouraging the movement of their senior manager. This might prove most burdensome to medium-size companies.

> "As technical changes are made more and more rapidly two things seem to be happening. On the one hand, companies which are established to exploit a particular technology have less time to achieve commercial success before new technical advances cause the products to become outmoded. On the other hand, each new technical development offers another opportunity to defectors from established companies to start their own new ventures. It is, therefore, becoming, perhaps, even more difficult for medium sized companies to make the important transition to a larger size and a greater

---

20 Clark, R. *A View of US Venture Capital,* The Technical Change Centre, Mimeo, London, 1983.

range of products. Just as their existing markets are becoming more competitive, their chances of developing new ones are diminished by the loss of their best people[21]."

Financial questions, then, can be of variable significance but they are part of the process by which innovation is managed. The case studies presented here illustrate a selection of innovations with different costs in which cost and providing finance play a more or less significant role. Lack of adequate finance was never suggested as having stopped innovation, although lack of funds had altered the receptivity of the corporate culture to risk-taking. A successful company must manage to grapple with such problems. For a less successful company which has slipped back, recovery can be problematic.

The importance of financial difficulties as a constraint on innovation as reported in most studies is much the same as it was when diagnosed by Carter and Williams in one of the earliest studies. "Out of the 138 firms for which information exists, lack of finance was a definite hindrance to the full use of science and technology in 19, while in 119 no definite hindrance from this cause could be identified". The authors thus concluded "... that shortage of money is an occasional hindrance to the full use of science and technology, when taken in conjunction with other features or attitudes of the firm[22]". Even in a buoyant venture capital market the capital is only part of the solution. "The recipe for a successful new venture, nearly all venture capitalists agree, is good management ... A company may fail even though it begins with a good product; but a good management, in the right market, can overcome all sorts of deficiencies in the goods the company is selling[23]."

The conclusion of this report would be that the first step on the road to recovery would be an analysis of the management weaknesses which led to the initial failure and its rectification. If this had been completely beyond the company's control a move to strengthen those arms of management most needed to combat the existing adverse circumstances would be advised. Adequate resources are vital to successful innovation. In particular countries, or at particular times,

---

21 Clark, 1983, Op.cit.

22 Carter and Williams, 1957, Op.cit.

23 Clark, R., 1983, Op. cit. See also N. Macrae, "Small Business Needs Marketing as Well as Money", *Finance for Industry Occasional Paper 2,* London, 1983.

or in particular industrial sectors such resources may be difficult to find; but finding such resources, using them well and earning an adequate return on them is a central part of the management task.

IV. The Relationship Between R & D and Innovation

It is generally agreed that it is vital for competitive success for an industry to be innovative. Innovative activity is made up of many strands, but R & D activities seem to be a major component in the total[24].

EIRMA STUDY

A recent study by the European Industrial Research Management Association[25] (EIRMA) attempted to answer the question "What would have happened to innovation without R & D?" by analysing 32 cases of innovation in industrial organisations. The sample, it was suggested, was somewhat biased towards technology-based innovations, since it came from people working within the research and development area. The study was also exclusively of large companies having their own R & D organisations and possibly was not representative of the situation in smaller companies. Case studies were examined against a group of stimuli to innovation which were considered to be significant by the EIRMA working group. Stimuli included the following needs:

(1) to preserve and promote existing business,
(2) to diversify,
(3) to respond to governmental and societal pressures, and
(4) to respond to scarcity of resources.

The group found that it could generally be concluded that the stimuli for innovation lie principally in the preservation and protection of existing business. Diversification can also be an important stimulus, but put into practice diversification is more often related to the extension of existing mainstream business, rather than development of entirely new lines of operation. Current innovations were more likely to owe their genesis to pressures to preserve and improve the environment and to the need to conserve energy and raw materials than had been the case in the past.

---

24 Pavitt, 1980, Op.cit.

25 "The Role of Industrial R&D in the Innovation Process," European Industrial Research Management Association Working Group 27, *Proceedings of the EIRMA Annual Conference "From Research to Reality"*, Paris, 1981.

An attempt was made to analyse the source of ideas for innovation. The internal sources suggested were research and development, engineering and manufacturing, marketing and information flow, and spirit. External sources suggested were the scientific environment, competition, customers and suppliers, and the government. R & D was found to be by far the most important source of ideas and was followed by marketing, engineering and manufacturing. The external sources were considered less important; the scientific environment, customers and suppliers received most citations.

It was concluded that internal innovation 'push' was more important than external innovation 'pull'. The report also suggested that the major factors in encouraging innovation were in-house scientific and technical capability and the availability of applications know-how, followed by the involvement of R & D at all stages, and effective communication and co-operation. Communication between R & D and marketing was stressed as of particular importance.

Although the exercise was a study of successful innovations, an attempt was made to analyse some factors impending innovation. The only factor within the organisation in general which was considered of significance was "imbalance between technical and commercial capability". Poor communication was seen as of some significance as were problems concerned with the transfer of results from R & D. Externally the most significant factor was customer inertia.

The EIRMA working group compared their results with those presented in the literature, and found a great degree of consistency. They concluded that their analysis of case studies had allowed the identification of some rules with respect to the role of R & D in industrial innovation. Matching the R & D expenditure to the company budget, they suggest, belongs to managerial art in the same way as policy making does. Many variations were thought appropriate for making the best use of research and development, depending on the branch of industry, the strategy of the company, market opportunities, the economic environment and so on.

The most important role of industrial R & D in the innovation process, the report suggests, is its capability to contribute effectively to a company's innovation potential. Successful innovation is a measure of the effectiveness of R & D, and a way of checking its usefulness. Thus the overall role of R & D in the innovation process can be used as a measure of how well its potential for innovation is used by the company. The importance of the role of R & D in the innovation process varies. In general, the more the innovation is a radical one with respect to the state-of-the-art in the market-place, the more important is the role of R&D. Since the support of top management is crucial in radical innovation, such an innovation is more likely to occur in an R&D orientated company. The report, although it stresses the importance of the quality of the R&D professionals and their role, also suggests that a project champion is often essential in pushing forward an innovation.

However, even in the most successful cases of innovation, R&D is only part of the effort, and the success of its contribution can depend to a large extent on how good the relationships are between R & D and other departments. If the communications links are good, and extend up into top management, the innovation process is likely to be more effective.

The view and suggestions in the EIRMA study are largely echoed in this report. The original work on the organisation of R&D in major British companies which has been used and expanded in this BNAC study was completed in 1982[26].

That study of the effect of recession on industrial research and development covered 35 companies ranging in size from 5,000 employees to over 100,000, seven industrial sectors were covered: electrical/electronics, food processing, chemicals and pharmaceuticals, engineering and vehicles, mining and metals, oil, and a residual category of other companies. The effects of the recession upon the R & D effort in the companies was analysed both in terms of its impact on particular sectors, and in terms of its effect on particular project areas. The study did not deal just with the immediate, more predictable effects of recession, but with the whole context in which the companies operated over the recessionary period.

THE ELECTRICAL/ELECTRONICS INDUSTRY

The companies in the electrical/electronics and pharmaceutical sectors which took part in the initial study were again approached for further information about the process of innovation for this study. Two American and one Canadian electrical/electronics companies and two American pharmaceutical companies also participated in the

26 Swords-Isherwood, 1983, Op.cit.

work. This work on R & D and innovation again revealed the marked importance of sectoral differences.

Originally the UK electronics sector had felt the effects of the recession in relatively minor ways. This was consistent with the Confederation of British Industry report which showed R&D spending in electronics in the UK up by 14% between 1978 and 1981[27]. Generally it was felt that growth in electronics would have been faster without the recession, although this might have been limited by a shortage of suitable technical labour. Changes in government policy were seen as the most significant happening of recent years. Table 1 shows industry R & D spending in OECD countries in 1977.

Table 1

THE VALUE OF ELECTRICAL AND ELECTRONIC INDUSTRY R&D FINANCED AND PERFORMED IN 1977

£ (million)

	UK	*France*	*West Germany*	*Japan*	*US*
Electrical and Electronic	£209m	297	568	794	1436

Source: OECD.

A Canadian electronics company was forced into a more thrusting role by the need to compete effectively in the US, since it planned to grow to multi-national size and the Canadian market alone could not have supported such growth. The basis of its competition was better technology.

The rate of change in electronics, it was suggested, is now so rapid that change is the norm, and stability is abnormal, although designing for change is almost a contradiction in terms. The recession had made the company "think twice" rather than alter the outcome of decisions. Difficulties in innovation arose from imperfect communication and the difficulties inherent in blending individuality and team work. "The company must not become comfortable, comfort means

27 *R&D in Recession Too,* Confederation of British Industry, London, 1983.

no innovation. If you get too good you must set up internal competition. You need to confront the unknown on a continual basis and that is hard to do."

The company provides maximum access to computing always. It has concluded that the innovative process is manageable and is concerned with the commitment which is needed at the top, from the top 10 executives and from R & D. R & D personnel must also be part of the top 10 team on equal terms.

The company feels it is useful to keep the age ratio low in R & D by moving people out into other functions. Youth, it was thought, had made R & D in the company. Now commitment and enthusiasm are high, and the knowledge base is up to date. The company has so far spun off 32 companies, and this is encouraged. It regards itself as the leader in market-driven R & D, the leader in micro-electronics, the leader in design systems and the leader in service. Equally important is the close relationship between manufacturing and R & D.

All of these strategies require talent within the company and so it is always looking for talent. Management rely on government support and university tie-ins in this area. The optimum laboratory size in the company is 500, of whom two to three hundred are technical people. There have been changes in the make-up of the laboratories with an increasing expansion of software. The company view is that to be successful in innovation, decision-making processes must be shortened. Inevitably developments are going to take time, but the company must always get things to the market before competitors do. This is difficult, but necessary, and the company have achieved timely competitive success.

On any project two-thirds of the R & D work is done in the central laboratory and one-third in the plant. This ratio was felt to strengthen the relationship between R & D and manufacturing.

This positive aggressive attitude was most marked in the Canadian electronics company, although to achieve it in Canada an "educational process" has been required. The relationship of British companies in this sector to the Ministry of Defence leads to market distortions which many suggest have adverse consequences for competitive behaviour.

The two American companies visited were insistent on the need to keep the interest of the corporation centrally in mind. For one respondent a major driving factor now was increasing international competition, particularly from Japan. The company felt a great sense of urgency to stay competitive and the drive in the innovation process came from R & D itself. The company is now more than ever alert to the market-place and attempts to improve its market sensitivity to counteract the technology push within it. The work in the company is multi-disciplined and so are the management team, although most have a technological background. The company regards itself as innovative and sees innovation as a process from research to the market-place. Ideas in the company have been plentiful but finances have been more restricted and this has influenced management's willingness to take risks. Efforts are being made to overcome this, and there is a new concentration of effort on high technology areas of company strength and excellence.

The second US company respondent contrasted his old, less innovative, bottom line oriented, well established company with the "darlings of Wall Street"—young and innovative firms with a high price-to-earnings ratio and little concern for a bottom line. For this respondent innovations required a combination of technical and entrepreneurial skills and a persistence in the face of uncertainty, risk and change—"the very things that humans are socially opposed to".

Too much technology push and not enough market pull was a problem in the company, but a programme had been instituted to help overcome this. For this respondent, the strength of the company's R & D function lay in its high technology support for the divisions. Its weakness lay in creating new commercial products. This was seen to some extent as a reflection of corporate strategy, which the respondent regarded as inadequate to enter new high growth businesses. Changes would be needed, he felt, both in R&D and in perceptions in the rest of the company.

THE PHARMACEUTICAL INDUSTRY

In the original study, UK chemical sector R & D spending had grown by 4% between 1978 and 1981, although this growth rate was lower than that of the previous six years. The UK chemical companies had maintained overall spending and effort, but had been subject to major internal company changes which were regarded as of more significance than recessionary pressures. Recession was said to have focussed attention on purposes and to have led to efforts to identify key areas where there were both changes and possibilities for growth

In pharmaceutical companies within the sector, research was said to have played a vital part in the success of the British pharmaceutical industry, which in 1979 ranked fourth amongst world exporters in pharmaceutical products (see Table 2).

Table 2

WORLD TRADE IN PHARMACEUTICALS 1979

US$ (million)

	Exports	*Imports*	*Balance of Trade*
Switzerland	$1,506m	$340m	$1,166m
USA	1,699	712	987
UK	1,358	495	863
West Germany ..	1,980	1,139	841

Source: UN Yearbook of International Trade Statistics.

Such research is essentially high risk and long-term. Costs of R & D were said to have risen rapidly, more than three-fold between 1968 and 1978. R & D expenditure in pharmaceuticals in the UK reached £208 million in 1980 and has grown at a faster rate than most other sectors of manufacturing industry. Recession had had very little effect on this picture. Most companies felt their success was sufficient to allow for continued growth through the recession. Any stagnation was felt to be particularly damaging, since getting new ideas and new blood into a research-based industry is of the utmost importance.

This general satisfaction with R & D and the future of innovation in UK pharmaceutical firms might be misplaced, if a recent study of the US pharmaceutical industry has any lessons for the UK[28]. The US pharmaceutical industry, it is suggested, has maintained an image of immunity from deterioration of its competitive position which is probably untrue. The data compiled on the industry indicated a clear relative deterioration in the foundation of the pharmaceutical

28 *The Competitive Status of the US Pharmaceutical Industry,* National Research Council, National Academy Press, Washington D.C., 1983.

competitive position, that is, in the research effort. This discrepancy between image and reality results in part from the time lapse between strategic decisions by ethical drugs firms, and the impact of these decisions on the market. The report suggests that the US position in the pharmaceutical innovation process "indicates a marked drop for the US presence in world pharmaceutical markets around 1960, followed by stability in the US share of new drug introductions and sales (outputs of the innovation process). In contrast, the US share of R & D expenditures and drugs entering clinical trials (inputs of the innovative process) has continued to decline, strongly suggesting an eventual further decline in US shares of introductions, sales and exports[29]. Furthermore the report notes that the steady decline in the American share of world pharmaceutical effort is more severe than the changes in world R & D share for other American industries.

This less than optimistic picture of the industry is counter-balanced by the very positive approach to research, development and innovation in the two US companies visited for this study. One of the two US pharmaceutical companies visited had recently increased its research effort significantly and had hired a further 35 PhDs. The research effort had been increased annually by about 20% over the last few years but the company did not regard itself as a leader in the field. The pharmaceutical industry, it was suggested, is generally recession proof, and R & D in the company had grown during the period of the recession, although cost consciousness had also grown. The company was now making a sustained commitment to innovation, and attempts to integrate the R & D function more closely with the rest of the business, and to increase the role of marketing in innovation, formed part of this effort. The respondent in the second US pharmaceutical company visited suggested that only industry could afford the investment necessary to develop a new drug; thus very few discoveries and no development come from outside the industry. New products were seen as the lifeblood of the company, and a steady stream of them is needed, but this requires great investment in a very competitive industrial environment. R & D in this company had increased in real terms despite the recession, and even if the industry were not relatively resilient to recession, R & D would always be the last to be cut. This respondent made a distinction between innovation and creativity and felt that an innovative company required the right attitude in top

29 National Research Council, 1983, Op.cit.

management, a willingness to commit adequate funds to innovation, a desire to succeed, and management skills. One of the UK respondents stated succinctly, "Innovation requires experience, confidence and internal standards".

The pharmaceutical research in this company had been restructured and put into one new division operating as a world-wide research division with six major areas of research. Previously R & D had been attached to the business units, but the new central function had allowed the company to attract new talent and to "achieve a critical research mass". The selection and retention of high calibre researchers was seen as vital. For this company "innovation is competing with yourself, a constant renewal, a continuous process". The company needs to create "a climate that welcomes new ideas and encourages internal debate. Innovation is not just the discovery of a new product but continuous renovation within the company".

There was very great agreement amongst all the pharmaceutical respondents in the study. For all companies in the pharmaceutical industry in the US and the UK innovation was necessary for competitive success, and research played a central and major role in the innovation process. All were agreed on the difficulties and expense of the development stage in pharmaceuticals. The influence of government through regulatory practices was highly significant in the UK and North America and changes in such regulations could defer or defeat the development of an innovation.

An analysis of issues which arose from the TCC and BNAC studies showed that comparisons of R & D effort and expenditure without clearly agreed definitions could be misleading. Such agreement on definitions had not yet been achieved. However defined, an in-house R & D effort was considered by all the companies to be essential for success. Companies in the sample had re-organised their R & D structure as a direct consequence of recession, often to overcome problems of isolation of the research function, and to integrate the R & D effort with company needs and with business planning. The recession had often resulted in a new cost consciousness about R & D expenditure and had led to renewed attempts to improve methods of R & D costing and justification.

Overall, more effort was being directed to research concerned with increasing productivity as well as to research for new products. Calculating the effectiveness of the R & D effort was generally regarded as difficult, although not all respondents were concerned to attempt it.

In both electronics and pharmaceuticals recognition of the important role of research, development and innovation in competitive success is universal. Even here, research and development, although necessary, are not sufficient for successful innovation. "Research and Development are integral parts of the innovation process, but they make no economic contribution unless the knowledge generated is integrated into the production process or manifested in a new or improved product that is sold on the market[30]."

The importance of other business functions, in these sectors marketing particularly, was referred to by all respondents. These two sectors, electrical/electronics and pharmaceuticals, might be taken as one end of a continuum. Here, research is vital in innovation but still only part of the picture. At the other extreme, the role of research, or perhaps, in this case, more properly development or technical expertise, would be small and the importance of other functions greater. The conclusion to the case studies which are discussed later in Part I of the report provides examples of this variability.

30 Schacht, 1983, Op.cit.

V. A Union View

This report concerns innovation and the problems of managing innovation, but innovation involves not just managers but the whole workforce. None of the case study respondents had experienced any industrial relations problems concerned with innovations within their companies, nor had the process of innovation been inhibited by resistance on the shop floor. Indeed management resistance to change had proved more problematic. Although innovation might be inevitable to ensure competitiveness and economic growth, the cost of change often falls disproportionately on certain individuals. Many companies in this study were at pains to ensure that this did not happen and provided guaranteed employment and retraining opportunities Some industrial sectors have suffered more than others and the responses of unions to such challenges have varied. The view outlined here illustrates a positive response to severe pressure. The case is taken from the apparel and textile industry.

Resistance to change, it was suggested, is universal and it is foolish to expect trade unionists to be less resistant than anyone else. This respondent's union had, however, become involved in a technical change and had recognised a need for its industry to modernise and to stay competitive. The union was involved in a joint labour, management, government project which was an unusual approach to the problem. This cooperation resulted from a study which had indicated an opportunity to exploit the market. The success of the project had now narrowed the differential between domestic and offshore prices and it has been a particular success in the one area of the unions interest.

The apparel and textile industry is a seasonal industry and until recently its fortunes have been very closely related to economic recessions. The union had experienced high unemployment in the last two recessions and was interested in job retention and job security. It was helped in implementing changes by the high employee turnover in its industry; thus the steady employees saw the long-term potential of agreements, whereas the shorter-term employees change fairly rapidly anyway. The industry is a piece-work industry and low paying. It has experienced geographical shifts in operation to the South where it has relied upon the labour of poor white and then southern black workers. Within the industry covered by the union, there is very little R & D in the apparel end of the business, although some companies have undertaken R & D mostly in engineering. Now \$5-6 millions (US) are spent on such R & D, which is substantial for the US industry and is a first step to catch up with Japanese, who spend \$40-50 millions in this area. The union is also interested in quality work circles which it hopes will help it face competition from Japan and from domestic competitors.

The union feels that new technology has actually been used to save jobs that would have gone to the Third World. Industry reorganisation is resulting in fewer larger companies, rather than many small ones. All these changes are easier in the menswear end of the business, where styling remains fairly constant. The union is now concerned to set up a retraining scheme, but there are difficulties in doing this because most of its members are not trainable for high technology jobs. Legislation on health, safety and related matters has also made the industry more competitive.

The union covers both textiles and apparel. Textiles are dominated by big companies where technology and innovation are advanced. The apparel business, the respondent suggested, is a totally different industry from any others dealt with in this report and is unusual because it involves innovation in many small companies. The burden of innovation very often falls on the union member and little is done to ease the technical transition for the individual worker. Obviously retraining has potential here and the respondent's union has negotiated retraining schemes and has won from management a more responsive attitude toward the needs of employees.

The union has met with resistances, from its own members particularly, for instance, from the more highly skilled because often their jobs are being de-skilled and consequently they face the threat of losing their jobs. All technical change alters the life and status of employees—those with high skills often have high status and, therefore, may lose both. "Nobody likes to accept the loss of their skill or their job", but the kind of developments this union has engaged in have, the respondent felt, protected, insofar as it was possible, its members' interests.

VI. Case Study Findings

THE INSURANCE INDUSTRY

The essential role of data processing, and its contribution to the insurance industry and its development, was heavily stressed by the companies involved in the insurance study. Although all the companies were expecting major data processing developments in the future, the nature of these advances was less clear. Reliance on IBM as the main equipment supplier was marked, as was the general level of satisfaction with available hardware. Software supply and development was more troublesome. Secrecy over innovative data processing developments was thought unimportant, even inappropriate, in the industry. The need for a supportive corporate climate which both protected and rewarded the innovator was recognised. Even within such a climate the innovation champion still had a significant and important role to play in bringing a good idea to practical fruition.

For the most part, funding for innovation was not regarded as a constraint. A clear trend was the move towards encouraging company personnel to access data processing facilities directly, and towards developing "user friendly" languages, but fears of increased error were inhibiting this trend. There was disagreement on the ideal background for personnel who should be involved in innovation decisions with some companies favouring data processing skills and others preferring an insurance bias.

The major difference between North American and UK companies lay in the acceptance or rejection of the label 'innovative'. UK companies claimed to be innovative and clearly related the term to efficiency, esteem and maintenance of market position. North American companies, on the other hand, favoured the 'fast follower' label. They feared that to be regarded as innovative could connote being seen both as putting technical advances in data processing before the business needs of the insurance industry, and as accepting avoidable risks. North American companies wished to be seen as advanced, but not to be involved in what they feared might be seen as the foolhardy vanguard.

Despite this difference of emphasis and the varied legislative environments in which they worked, all the companies can be regarded as highly innovative. Technical advances in data processing, seen initially as a means to increase efficiency and reduce costs, soon allowed opportunities for product innovation which would have been impossible without the former. Selling data processing and the perceived potential within the company of expanding this facet of the business, often resulted in the development of a greatly improved marketing ability and a new market orientation. The rewards of innovation and the costs of non-innovation were accelerated by rapidly increasing competition in the industry. Entrance into the insurance arena by other financial institutions, and the blurring of boundaries between insurance and other financial sectors, mean these pressures will become greater and the innovative importance of data processing in achieving competitive success can only increase.

THE PAPER INDUSTRY

There are opportunities for both product and process innovation in the paper industry. These could occur in the development of raw material substitutes, by achieving energy savings in new developments, and by adding greater value to an essentially low value product, but the pressures under which these opportunities must be taken are very considerable. The paper industry provides one of the clearest examples in the report of how difficult it is to provide easy formulae for successful innovation.

The cyclical nature of the paper industry is an immensely significant factor affecting its ability to innovate. These cycles reflect the cyclical nature of general economic trends. The paper industry is a basic mature industry, so variable financial factors such as exchange rates and interest rates are very significant in determining overall profitability. A company may be innovative, and potentially profitable, but an adverse change in exchange rates can wipe out this advantage very rapidly. An emphasis on cost saving in the face of competition is very important at the low added-value end of the market. The high added-value end of the market is also subject to immense competitive pressure.

Throughout the market the impetus to innovate is very strong and the need for innovation is great. This is a result partly of the competitive nature of the industry itself but also partly a result of substitution from outside the industry, from products and processes such as those arising from the continuing diffusion of information technology and plastics. At the low added-value end, however, companies are in delicate financial situations. Process innovations are necessary for success and for survival but if such developments cost money, while maintaining or reducing costs in the short-term is also essential, then taking an innovatory decision requires a considerable amount of confidence in its successful outcome. Unfortunately, many companies in the paper industry lack such assurance. Product innovation within the paper industry is slow, not particularly radical and produces innovations which are relatively quickly and easily followed by others. Companies, therefore, cannot gain an unassailable competitive lead in innovation. Even if companies have been innovating successfully in their products, they rapidly move into competition on costs.

It is easy to see how a company operating in such an environment could become over-concerned with cost. This can have serious implications for achieving the appropriate balance of management skills, and the necessary mix of management influence within a company, which seems to be vital in successful innovation. If cutting costs becomes the paramount interest, then the power of those most obviously concerned with cost cutting—that is, the financial and accounting people in the management team—can easily become too strong. The R & D, production engineering, or marketing voices which might be proposing an initially costly but ultimately rewarding innovation are likely to go unheeded.

The materials and resources cases illustrate well that too much emphasis on cost factors, however derived, tends to become more unbalanced by further apparently logical but eventually disastrous attempts to cut costs. Clearly, the degree of movement from the appropriate management balance which is possible differs between industries. In an R & D-led or dominated industry, like electronics or pharmaceuticals, it is quite difficult to stray to a position in which R & D has no significant voice, although other imbalances are possible. This is not the case in a basic industry like paper where profits may be small and threatened. The immediacy of the dangers or imbalance relate to the financial strength of the sector.

Other factors in a company's corporate culture or in its national situation which inhibit the development of a management team with skills balanced appropriately to compete in that industrial sector could also inhibit innovation and competitive success. Such imbalance can also be seen in the paper industry.

Company H in this study seemed to be becoming understandably, but perhaps excessively, concerned with the cost side of its enterprise. Company G, although recognising the vital nature of the cost element of the business, had attempted through the development of business teams to re-assert the importance of research and marketing functions in the company's strategy for competitive success. Company F like Company G had a major re-organisation, consequent on a difficult financial situation, but had also attempted to maintain its innovative capabilities. The appropriate corporate climate for innovation which involves a strong commitment to it, and a suitable business orientation, was considered by Company F and G to be fundamental to success in innovation. Company H on the other hand, stressed individual talent and the availability of money, rather than the development of an effective management team, as essential to the innovation process.

THE MINING AND METALS INDUSTRY

It could be argued that neither Company V nor R are really typical of the metals business since they both are in specialist metals areas and are both relatively large and strong. Perhaps they are not therefore truly representative; but all suppliers including Companies V and R are in the frustrating position of being dependent on others for a means to recover from the recession. It is very difficult for them to improve products since any changes are probably not acceptable to customers at such a time and, therefore, this leadership must come from the users. Although metals companies are supplying across industrial sectors, they are mainly supplying the engineering and building sectors which were and are particularly badly affected by the recession. For the metals industries there is a need to operate near capacity levels to make money, so plant loading is a vital factor. When R & D expense is carried on top of an active sector, the costs are not too difficult to bear, but such costs on a more limited production can become too expensive to undertake. Any demands from customers to provide new materials or better products are very costly. Such dilemmas are less extant if the metals company is involved in a final product, for example cables. Exacerbating these trends, there has been a move towards the use of less material in a product. The trend now is towards limited life products, often of different materials, in particular, plastic. The material sector can, however, come back to prominence very quickly when industrial growth accelerates; therefore, it is necessary to be in a position to maintain R & D effort in order to take advantage of any upturn when it occurs.

Company V has always been regarded in the industry as one of the most intellectually powerful companies. It has now had to adjust its sights. There is still a very real commitment in Company V to innovation and new products. Technically the company is well equipped to analyse and react to likely changes in the business environment. Company V has shed its innovatory work in batteries at some cost, while Company R has increased its work in these areas of conventional lead acid batteries and expects this to be profitable in the future. Despite dramatic external changes, Company R has steered an even course, seeking to be known as efficient and highly competitive rather than innovatory. Company R has been concerned with improving the efficiency of processes and searching for new markets. It has a competitive edge in zinc and lead batteries but the markets for these are changing. For Company V investment in innovatory research for new applications for its products is rated of higher importance than it is in Company R.

THE OIL INDUSTRY

Although maintaining their positions as some of the largest companies in the USA and the UK, the oil companies are experiencing relatively hard times. Companies O, P and Q have poured considerable amounts of money into R & D. One US company has taken a decision to concentrate on its primary market and look for innovations in more limited fields. It has been concerned with a major innovation for the production of oil from shale. The other US company, like the UK company, has re-organised. One of the major objectives of this re-organisation was a decision to move into more speculative kinds of research. Company O is actively researching in the areas of biotechnology and coal gasification but with "business objectives always in mind".

Companies O, P and Q were concerned with secrecy in certain circumstances. All perceived secrecy as necessary in obtaining a leading edge which they view as significant in maintaining returns on investment.

All the oil company respondents saw commitment to R & D even during recession as vital to innovation. The commitment must be financial but must also be reflected in attitudes throughout the company. Market and cost considerations were also seen as important to innovation. In Company Q the innovation champion was seen as having a vital role.

THE MACHINE TOOL INDUSTRY

Rapid and dramatic technical changes in the machine tool industry have required a significant change in the skill composition of the workforce. This includes increased understanding by those in management of the implications of new technology for corporate decision-making and planning. Both Companies S and T have tried to face up to these new pressures. Company S has invested heavily in micro-processor based equipment and Computer Aided Technology (CAD) and is now researching laser technology and employing more university graduates. The company, although engineering oriented, recognises the importance of the market and monitors changes in it, in customer needs, and in competitors' actions. Whether other management functions effectively balance technical engineering capacity in Company S is open to question, but the possibility of imbalance has beed recognised. Company T, although hard hit by the recession, has continued its attempts to be innovative and is prepared to face up to its Japanese competitors. To this end the company spends 5% to 10% of turnover on R & D (a high proportion for the industry) and, although more concerned with progressive iterations than with radical innovations, is investing heavily in new technological opportunities. As in Company S, market sensitivity was considered of importance to innovative success. The need to monitor technical changes outside the industry, for example in electronics, was stressed. In both companies innovation was seen as requiring talented individuals but more importantly serious company commitment. Given the nature of the competitive challenge facing Company T, its commitment will have to be total.

THE BIOTECHNOLOGY INDUSTRY

The primary difference in the development of industrial biotechnology in the UK and the USA is that the UK industry has been dominated by the large diverse firms which were already well established in other fields. The US industry has developed with a large number of small companies, many of them surviving on venture capital.

The US business community was more ready to invest in biotechnology at an early stage, while UK financiers were slower to respond, even though the UK had a sound basis in research in the biological sciences in the early days. Since the publication of the Spinks' report, there has been a move toward increasing public and private finance for research into biotechnology in the UK, and the research effort is now more coordinated.

In both countries, it was suggested that the interest in biotechnology was more likely to be directed towards medical applications and pharmaceutical companies. This it was thought might result in a shortage of good scientists available in the plant science field, the area of interest to both of the companies in this study.

In both the UK and the USA the general economic climate was seen to have adversely affected biotechnological innovation, as had the lack of thoroughness of commitment to biotechnology.

THE MEDICAL IMAGING INDUSTRY

Both companies in this sector had grown rapidly in recent years, although in the case of Company K growth was only achieved after a balance of good technical skills and business management was obtained in 1973. Previously, the over emphasis on technical capability and the underdevelopment of business and marketing capacity had inhibited competitive success. Establishing an appropriate management balance and developing a group strategy which minimised the chances of future imbalance has led to a dramatic improvement in company prospects.

The danger of becoming too concerned with technical innovation and achievement at the expense of commercial viability was recognized in both Company K and L as a major pitfall to be avoided.

Although stressing the importance of being at the technological forefront of the industry, both companies were fully aware of market requirements and their importance. For them innovation required both technological push and market pull. The strategy followed by both companies of setting up new operating companies for each new product range developed is designed to keep them in touch with their customers and to ensure their responsiveness to the market.

Both Company K and L emphasised the need to recruit and hold good people who are committed to the company and to innovation in order to achieve technical and commercial superiority. Good people, it was felt, attract more good people and, for Company L, employee stock (share) options were an important factor in recruitment. Company K also offered shares to its older employees and felt this cemented company loyalty.

Company L has in the last year suffered a considerable business setback and has withdrawn from one of its three main areas of business completely. At the moment the company seems to be surviving this major loss.

THE ELECTRICAL/ELECTRONICS INDUSTRY

Both Company N and M are small high technology companies which have grown rapidly. The business environment in which they function demands that they are innovative both in technology and in business methods. Company M, however, sees itself as having a technology looking for an application, while Company N suggests its innovation is customer directed. This difference in emphasis follows in this case the international pattern often suggested, with the UK company emphasising technology and the US company the importance of the market. Both companies acknowledge the need for an aggressive approach in both areas. Company M was founded in response to a growing market need. Company N and Company M both work with very short product life cycles, work at great speed, and are under continuous pressure to innovate. Their size also means their profitability and survival depend on the new, rather than the 'me too', innovation.

Both companies follow a pattern of finding a few creative people and supporting them adequately. Small company size was seen as important to this strategy. Company N favoured dealing with smaller rather than large firms in the interests of good communications.

The Company M respondent stressed the open nature of the industry and the importance of cooperative ventures. He regarded over emphasis on secrecy as a British mistake. Company N on the other hand saw itself rushing for copyright on devices and consequently rated secrecy as rather more important.

Both companies experienced problems with the supply of qualified personnel of the right calibre. This is general in the electronics industry.

Successful innovation was seen in both companies as a product of a rapid response by creative people, supported by an adaptive management team committed attitudinally and financially to its importance.

VII. General Conclusions

In a cross-national study, conclusions which emphasise national differences in approach to innovation might be expected. There is evidence to suggest that at an aggregate level, the US is more innovative than Canada and the UK[31]. Since the role of management in innovation is suggested here as crucial, and the technical quality of management differs in these three countries, certain consequences could be expected to follow from this[32]. Many of the companies taking part in this study are large successful multinationals; others are small highly innovative companies in internationally competitive markets. Many participants are members of the British-North American Committee and so have evidenced their international orientation. Thus it might be argued that the nature of this study, and of the companies participating in it, reduces the likelihood of significant international differences emerging. All the participating companies are interested in innovation and its problems.

National differences in the objective circumstances in which participating companies were attempting to be innovative clearly existed but seemed to play a role only in helping or hindering their progress. There were differences: biotechnology had for the most part developed in large companies in the UK and new small companies in the US; in microelectronics the UK company was more technology oriented and the US company more concerned with the market. This had also been the case in the early days of the UK medical imaging company but of late it was the US company whose misreading of the market had caused business problems. In the insurance industry UK companies were more keen to be labelled innovative than their North American counterparts, although their courses of action did not differ greatly. Nevertheless, such differences as emerged were less marked than a reading of the literature might have suggested and seemed to be of limited significance.

Industrial sector differences, however, are of immense importance. Large multinational companies and small high-technology companies in rapidly changing markets compete internationally. National factors which foster or obstruct innovation, although they clearly exist, have

---

31 Schott, 1981, Op.cit.

32 N. Swords-Isherwood, "British Management Compared", in Pavitt, 1980, Op. cit.

to be overcome at the level of the firm, rather than used as excuses for inaction.

Some new technologists transcend sectors and indeed blurr sectoral boundaries and create new sectors. Attempts to monitor and predict the impact of such technologies across the full spread of any economy, rather than by sector, tend to disguise their significance. The orientation of participating companies was transnational, but sectoral. The opportunities for innovation and the constraints upon it were often similar in different countries, but different in different industrial sectors.

The sectoral differences attributed here might be thought to arise from differences in levels of technology. Nystrom makes a distinction between positional and innovative companies[33]. Positional companies operate in stable environments, eschew change and adopt a reactive attitude to market requirements. Innovative companies, on the other hand, accept that the future is uncertain. Many companies of course fall in the middle of this spectrum. One must ask whether the notion that any industrial environment remains stable in times of rapid technological change and increased competition can be accepted. Companies which accept this unchanging view of their technology often find that they have miscalculated. The machine tool industry provides graphic examples of this[34].

The emphasis on the importance of the level of technology is also stressed by Parker in his study of the management of innovation in eight small to medium-size UK firms[35]. He concluded that "Successful companies which are engaged upon the manufacture of highly technical products are characterized by their employment of highly qualified staff who possess special aptitudes for solving problems, by their aggressive research and development tactics, by an organic organisational structure, and by reliance upon the production of relatively small numbers of high value products." Low technology firms would differ in most respects and their innovation would be evolutionary and aimed at low cost manufacture. Parker suggests that " . . . survival of a company can be threatened through neglect of a single management activity . . . successful innovation will only be

33 Nystrom, H. *Creativity and Innovation,* Wiley, Chichester, 1979.

34 See Sciberras, E. and Payne, B. *Technical Change and International Competitiveness—A Study of the Machine Tool Industry,* The Technical Change Centre, London, 1983.

35 Parker, R. C. *The Management of Innovation,* Wiley, Chichester, 1982.

accomplished when every management activity is carried out in a manner appropriate to the level of technology at which the company operates[36].". He goes on to suggest that management requires breadth.

The level of technology then, dictates the appropriate management structure and the work of Parker, like that of Nystrom, implies a certain technological rigidity. Although entirely reasonable in stable conditions, periods of rapid technical change might somewhat weaken the analysis.

A level of technology with its appropriate management might cut across sectional boundaries. Within a firm, management appropriate to a low-technology level might be ill-equipped to see the possible opportunities offered by a new technological innovation, which could transform both the level of technology thought necessary in the sector and the competitive position of the firm. Only by looking at best practice in its sector and by being prepared for innovation when it occurs can a company achieve competitive success.

Logically, the management of innovation poses problems. It involves gaining commitment and co-operation both from people whose strength lie in their individual talents, and from those with corporate excellence. Being an innovative company demands continuing adaptation to change in an environment which routinizes[37].

Aspects of organisational structure effective in stable conditions can act as obstacles to innovation. It is difficult not to sit back after a major innovation which has often required considerable restructuring within the company. The amount of restructuring and reorganisation in innovative companies in the study is marked, and such change cannot be regarded as a one-off. Technical innovations, and a company commitment to seek such innovations in the long-term, require appropriate management talents and structures to initiate them, and a further acceptance that innovatory changes in management may then be required again.

The role of managers in the innovation process may then prove critical[38]. The rate of diffusion of innovation, which it is suggested is relatively slow in the UK, will also be affected by the quality of

---

36 Parker, 1982, Op.cit.

37 Burns, T. and Stalker, G. M. *The Management of Innovation,* London, 1961.

38 Schott, 1981, Op.cit.

management[39]. Many studies of innovation and technical change stress the importance of the role of management and the need to achieve a balance of management functions. Throughout the case studies company respondents spoke of the need for a supportive corporate culture, and a company commitment to innovation which pervades the company and comes from top management. This balance is not just between R & D marketing and top management, although much time and effort is spent within companies on improving this relationship; it also must include product engineering and manufacturing as well as planning and finance. The inclusion or exclusion of particular functions can prove decisive.

The need is for a balance of managerial visibility and power, an appropriate mix of management functions. The problem is not solved by asking "Are we weak in technology? Are we weak in marketing?" A solution can emerge only when the whole business process and all the interconnections between functions are seen as part of a total picture. Many studies have come to these conclusions, and the case studies in this report add confirmation. A company that is unbeatable in engineering cannot capitalize on this excellence unless costing is right, marketing is right, sales are right, and so on. Japanese companies pose such a competitive threat precisely because they take this total view of the enterprise. Such a transformation cannot happen overnight but it can be achieved.

Having arrived at a list of necessary functions, the next step is to ensure that the functions have some sort of effective power. One of the criticisms of management in the UK that is common at present is that there are too many powerful accountants. If the final decision on a project is made by an accountant, it is going to be quite a different decision from a decision made by an engineer. What this report is suggesting is that in all probability neither of those decisions would be the right one. Either functional experts must cooperate in reaching any decision or it must be reached by a process of 'creative discord', but the decision will be wrong unless all the participants have real power in the debate. This is not to propose general guidelines for the management of innovation[40]. Such guidelines often assume a pre-existing correct balance of power, co-operation and communication, and propose a

---

39 Nasbeth, L. and Ray, G. F. *The Diffusion of New Industrial Processes,* Cambridge University Press, Cambridge, 1974.

40 Parker, R. C. *Guidelines for Product Innovation,* British Institute of Management Foundation, London, 1980.

consensus view of executive teams which, if it does not exist already, might be impossible to achieve and would make impossible demands on the structure of the organisation. The most productive possibility in Britain and North America is agreement through positive conflict. Prescriptions for breadth of management vision and give-and-take in the management team are too vague, and often based on unreal assumptions about behaviour in corporations. The technical know-how attached to existing functional divisions is vital.

Although this might not be the case in an ideal world or in totally different institutional and corporate environments, it is unrealistic to aim to emulate a model of management which is foreign to existing forms of organistaion. Thus it is arguable that although the Japanese model might be said to incorporate notions of management balance most perfectly, other models can be equally effective[41]. Many theorists would disagree with this conclusion.

For Dore what makes an organisation innovative is partly the cultural climate in which it operates and partly its internal structure[42]. He contrasts two models of business organisation, the market model and the community model, and argues that the latter, which is closer to that found in Japanese firms has advantages.

In the community model "the major purposes of the organisation are purposes to which all members of the organisation subscribe, and they can be expected to allow these collective purposes to supersede to some degree their own private purposes". The author argues further, "The market type is one which seeks to recreate the principles of market interaction—the stand-off bargained contract—within the organisation itself".

The models are absolutely clear but it is highly debatable whether they do in truth reflect Japanese and North American business organisation. It is as unlikely that "individual material self-interest" is entirely dominant in the US as it is that commitment based on the unity of interest can be believed to be the norm in Japan.

Japanese society is extremely competitive and not notably tolerant of deviations or failure; thus the appearance of commitment might be judicious just as the appearance of dissent and acceptance of

---

41 N. Swords-Isherwood, "Management Styles and Economic Success—Some Examples from the Television Industry", *Management Research News,* Vol. 5, No. 4, 1982.

42 Dore, R. *The Social Sources of the Will to Innovate.* Papers in Science, Technology and Public Policy published jointly by Imperial College, The Science Policy Research Unit and The Technical Change Centre, London, 1983.

conflict can serve useful purposes in the US. Disagreement and conflicts of interest can have positive results particularly in the areas of innovation where the search is for something new. Open disagreement does not imply lack of identification with the enterprise and its interest, just as apparent cooperation cannot rule out the possibilities of real hidden conflicts.

If the importance of the role of management in the innovation process is accepted, and the crucial nature of an appropriate management balance established, what is the appropriate management balance? Parker suggests that this differs with the level of technology and that the technical quality of management can vary in line with the level of technological sophistication[43]. It is proposed here that technical quality (possibly measured by numbers of graduates) must be high, whatever the level of technological sophistication, if innovative opportunities are to be seized. The appropriate management balance does relate to the level of technology, but is primarily derived from the industrial sector in which the company is located. In different industrial sectors different imbalances are likely to emerge. Further sectoral studies of innovation could allow the development of a checklist of imbalances likely in each sector and their causes. The present cases provide some interesting indicators. Examples include the importance of cost to material suppliers leading to over emphasis on cost and consequent lack of innovatory activity, which reinforces cost constraints; over emphasis on engineering with insufficient attention to marketing and costing in small high-technology companies, leading towards business failure; and lack of recognition of design and manufacturing possibilities in traditional manufacturing industry.

The British-North American Committee study could then be a step towards helping in the difficult process of ensuring successful innovation, although of course it remains impossible to provide a a recipe for guaranteed success.

---

43 Parker, 1982, Op.cit.

PART TWO:

INDUSTRY CASE STUDIES

Introduction

The case studies are presented fairly fully with, in each case, an introduction which attempts to highlight features in the industry environment which are of significance to the innovation process. The insurance industry is presented first and is used to illustrate the process of innovation in a service industry. The paper, mining and metals, and oil industries are grouped together in the report because, despite the very great differences in their financial outlooks, they have features in common with regard to innovation. For those who supply materials and resources to others, what constitutes an innovation is an even more difficult question to answer. Innovation may be concerned with pressing forward the use of materials and, therefore, there are limits inherent in what can be achieved by producers. Such companies, of course, can and do innovate on the process side. A machine tool industry case is presented as an example of innovation in a traditional manufacturing industry. Biotechnology, medical imaging, and micro-electronics cases are then presented and grouped into what might be termed a sunrise industry sector. The original case suggested by a BNAC member is always presented first.

I. A Service Industry

THE INSURANCE INDUSTRY

Overview: Computerisation in the Insurance Industry

The word innovation is not used as commonly in the context of service industries as it is in manufacturing. It is interesting then for this study to start with innovation in the service industry. Innovation in this context, must be defined as the use of advanced but not new technology outside the original setting in which it was devised. Possible dangers of innovation in insurance, as respondents make clear, lie in the possibility of forgetting the original business aims of the insurance industry and becoming too concerned with technical innovations in data processing. The benefits of successful and innovative use of data processing in insurance are to be seen both in vastly more efficient processes, and in the possible developments of new products.

Insurance was one of the first service industries to undergo mechanisation. The industry is concerned with large volumes of information and some highly routine processing. As early as the 1920s mechanical punch-card machinery was used and in the late 1950s there was a widespread adoption of mainframe computers, which continued in the ealy 1960s. Because insurance companies were centrally organised, they were very well-suited to these types of computers, which enabled them to handle a steadily increasing volume of business. At present, computing power and improved tele-communications allow possibilities of decentralisation of computing operations and this provides several possible ways forward for the industry.

There appears to be considerable scope for application of new technology in this industry. Stamford Research International has identified seven stages of technological development in insurance[44].

Stage I. Head office facilities using batch system. This initial stage entails the installation of a computer in the head office; all work is hand-carried to the computer.

Stage II. Head office facility using on-line system. In this stage, terminals in the head office are capable of data capture (putting data

44 Stamford Research International in *New Technology in Banking Insurance and Finance*, British Insurance and Finance Union, London, 1982.

into the computer), and enquiry (accessing the computer data base to get information in a video or hard-copy display).

Stage III. Remote offices have terminals on-line to head office computer. Use of the head office computer is extended to remote offices by installing terminals in them and connecting them by wire to the central computer. The remote offices can thus perform the functions performed by the head office. This may include issuing policies, printing policies in the remote offices as commanded by the head office (rather than mailing them from the head office).

Stage IV. Remote offices have minicomputers. This stage like Stage III extends computer capabilities by installing small computers in each of the remote offices to do some processing and by doing other processing in the central computer in the head office. Communication between the computers may be by wire, by paper, or by magnetic tapes or discs that are mailed or flown between the offices.

Stage V. Word processing facilities provided. This stage provides word processors that are typically small mini-computers with video display units and letter quality printers attached. The mini-computers are programmed to accept, edit, store, retrieve and format text for printing. They can be stand-alone devices, shared devices (one mini-computer with multiple terminals), or special terminals and software appended to the central computer. Usually no access to the business database of the central computer is provided.

Stage VI. Advanced functions utilised. This stage consolidates the systems offered up through Stage V. It requires a cohesive integration of the various systems and extends their capabilities further into the field and into business activities. Such extensions include:

(1) agent terminals to give agents direct access to computer systems for rate quotations, entering application data etc.;
(2) computer systems and terminals for job training; and
(3) word processing integrated with data processing to enable the combined use of text and numeric data bases.

Stage VII. Operating on a paperless office basis. This stage of progress is referred to as the "office of the future". It goes well beyond the use of today's simple word processors, focussing technology on the problems of the high cost of creating, storing, retrieving, processing and moving paperwork within the office. It encompasses:

(1) capability to capture documents in electronic form and to transmit and reproduce them;
(2) replacement of current file cabinets with electronic versions that utilise video terminals to create, store, search, sort, update and retrieve files; and
(3) terminals for sending and receiving intra-company correspondence (electronic mail).

It was suggested by BIFU (British Insurance and Finance Union) that most UK insurance companies would have reached at least Stage IV and that leading edge companies would have completed the first five or six stages.

A recent report from The Technical Change Centre (TCC) suggests that since the first stages of mechanisation, the industry as a whole seems to have followed a notably uniform succession of computer developments[45]. It is possible to identify a "natural trajectory" of technical innovation towards which all the companies visited for the study have tended to converge. This natural trajectory involves both hardware and software systems, and applications. The report describes the trajectory in terms of the stages of batch processing, on-line systems, distributed processing, and the integration of text and data. Individual companies in the study have progressed at different rates through successive stages in the trajectory, sometimes taking a lead and sometimes lagging behind. Indeed, companies who have had a leading edge with one generation of technology, although they may have enjoyed an initial advantage over their competitors, have often found subsequently that they have been left with an expensive and often disruptive conversion problem, which has been a considerable drain on resources, sometimes culminating in a period of lagging behind competitors. Those companies, which have delayed the introduction of new systems have often been able to catch up relatively easily and in some cases have even been able to leapfrog a generation of systems, thus enabling them to enjoy a temporary cost advantage, but again exposing them to the dangers of the vanguard user.

---

45 Barras, R. and Swann, J. *The Adoption and Impact of Information Technology in The UK Insurance Industry,* The Technical Change Centre, London, 1983.

Barras and Swann point to a gradual convergence upon a single supplier—IBM. All but one of their 12 sample companies used IBM equipment. This trend was also observable in the BNAC sample. "The result has been a reinforcement of the natural trajectory of computerization, to the extent that IBM's own product development strategy has become a crucial determinant of future technical progress in the industry[46].

In the early days of computing, with a central batch processing system, insurance companies often suffered from a division between the "them" of data processing and the "us" of the rest of the staff. Computers were seen to contribute little and to cost a great deal in terms of effort. With the advances of hardware, software and telecommunications allowing on-line users demands to be met, these attitudes have broken down and computers are now regarded as a valuable aid, which could and should enhance the service given to customers.

The next steps for the companies in both the Barras and Swann and BNAC studies were to use their networks for a wider range of fully inter-active applications, such as local data capture, direct update of files, and the local preparation of insurance, quotations and policies which can be produced on the spot for the customer.

Barras and Swann suggest these developments are being approached in two ways, either by on-line central teleprocessing or by local or distributed processing. Although they feel this would make little difference to the user, insofar as the development of the technology is concerned, it marks the divide in the path of the natural trajectory, at least for the present.

Barras and Swann suggest that, in practice, companies in their sample had adopted a half-way house solution, adopting distributed processing, but maintaining central data bases and using the branches for temporary local storage and for processing. Programmes and data are sent down the telephone lines from the central computer to branch machines, while details of the transactions are sent back up the line. The central computer is thus very much in control.

One of the final stages of development of data processing in insurance, outlined in both these studies, is the move towards integrated text and data processing. For Barras and Swann one reason for slow progress towards this goal could lie in the organisational structure of the companies. Traditionally there has

46 Barras and Swann, Op.cit.

been a distinct separation between management services and administrative services and, therefore, there has been a lack of co-ordination in their purchasing policies. Word processing has been treated as a separate activity from computing, and office automation has fallen between the two spheres of influence. A major technical constraint on text and data integration is the lack of satisfactory products to meet the requirements of the industry. At mainframe level there is a lack of compatibility between existing text and data processing systems. Here the companies' reliance on IBM means that their rate of technical progress depends largely upon IBM's strategies on software availability, price and, above all, compatibility and integration.

The TCC study concludes that the future of computerisation in the insurance industry will be towards distributed processing, but not necessarily distributed data bases. These network systems will allow integrated text and data processing locally and centrally, and the provision of personal computing and electronic mail.

The companies in the Barras and Swann study acquired their computing to improve their effectiveness or their efficiency. These continue to be the main reasons for the adoption of computer technology in the insurance industry. The main type of company business, long-term or short-term, explains which reason is pre-eminent. Long-term business covering pensions and life insurances has remained profitable and is still expanding. Improved effectiveness has here been a main motivation for adopting computer technology, to expand capacity and cope with growing demand in the future, and to assist directly in the design and implementation of new types of policies and services—that is, product innovation. In the static or declining parts of the business, the main motive for adopting computer technology has been to improve efficiency, that is to reduce unit costs, and maintain competitiveness. Computer systems have been used to make staff savings, to facilitate branch reorganisation, and to provide better information and control of the business. The choice of systems has been narrowing because each company is in effect locked into its mainframe supplier, since the cost of changing manufacturer is now great given investment in databases and staff experience with existing systems. As a result, it is now almost impossible for a large insurance company to change supplier. IBM has become the standard supplier for the industry, mainly because "it is the supplier company from whom a customer is least likely to be obliged to change[47].

The reasons for IBM's pre-eminence include its size and stability, the extent of its product ranges and the upward compatibility within its ranges. Major changes were faced by early vanguard users who had invested in the wrong hardware in the 1960s, and who were forced to change from their first supplier during the early and mid '70s. A second wave of change hit companies which at an earlier stage had invested heavily in their own software system development, to exploit the potential of their hardware. As relative costs changed and hardware became cheaper, non-standard software became a constraint on further developments, and a great cost burden to the company. A third wave of change could follow from the extension of the use of package application software now widely available and used in the USA.

Insurance companies experience few running problems, since they are large and mature computer users with in-house support and expertise available. Major problems are associated with the setting up of new systems, particularly when there is change in mainframe suppliers. A problem pertinent to insurance is the strategic technical problem of the threat of failure of the central processing system. Although long-term failures are unlikely, short-term breakdowns can cause serious problems. One possible solution is to duplicate the central processing system. In practice, few companies can do this, since running two parallel mainframes and databases would be prohibitively expensive. As a compromise, some companies have a second mainframe, primarily for other purposes, which can be used as a back-up machine. Others have duplicate and automatically switching parts. Most commonly, however, "users protect themselves through diversity rather than redundancy[48]" by distributing their processing around their branches between different parts of the business. Computerisation has led to substantial, but less than anticipated, job savings and productivity gains in insurance. Where markets have been expanding, improved effectiveness has been reflected in a rapidly growing volume of business handled by a slowly growing, or static number of staff. In other markets computers have allowed for increased efficiency through cuts in employment.

47 Barras and Swann, Op.cit.

48 Barras and Swann, Op.cit.

In terms of the trajectory of innovation, and of the seven stages of technological development, the companies which have taken part in the BNAC study on innovation in data processing in insurance are, indeed, innovative companies. They appear to be at the forefront in the use of technology, with one possible exception (Company C), which makes no claims to be innovative.

The following five case studies, Companies A and B in the UK, Companies C and D in Canada, and Company E in the USA, illustrate the variety of innovative uses of data processing, despite the constraints of reliance on one supplier, and accepting the notion of "likely trajectory" of innovation. Company A will be described rather more fully than the other companies in this section since it is the original case study.

Company A

Company A is one of Britain's leading mutual life offices, operating not only in the UK but also in the Republic of Ireland, Australia and through a subsidiary in Canada. The company was established in 1832, and in 1918 extended its scope to cover all members of the general public. It is a mutual life office with no shareholders. The company employs a staff of 1,250, has 500,000 policies in force with 70,000 new individual polices per annum. Long-term funds total £1,000 millions and the annual premium income is £150 millions. Company A suggested that it had been particularly innovative in its use of data processing and its claim appears to be borne out by the general regard with which it is held in the industry. The company is a medium-sized, specialist life insurance company, where insurances are marketed through 40 branch offices spread throughout the British Isles.

Originally the computer was seen as the answer to the need for speedier service and as a means of coping with the vastly increased amount of paper work which was being generated.

From the early 1970s, it was felt that the solution to company problems lay in telecommunications. The aim was for a "head office in every branch", and for "a paperless office". The objective was to connect directly every location in which business was done with a central computer and, therefore, to ensure that locations, however small, would have all the resources of the entire organisation available. The speed of service could then be fast if not instantaneous. This plan, initially felt to be ambitious, was helped by very rapid technical developments and dramatic falls in the cost of computers. The benefits gained from this computerisation were rather greater than had been envisaged. The company could provide a faster personal service, incurred lower costs and had far fewer mistakes and problems generally. The ability which the system gave to develop new products and to adapt old ones to meet new market demands had not been anticipated. The main system took 120 man years to develop, and four years elapsed time. The cost was £1 million at 1975 levels of expenditure. Company A now has a network of 300 terminals at 40 locations and the volume of traffic handled includes 2,000 quotations, 300 new policies and 90,000 messages daily with a response time of 95% in three seconds or less and availability of 98% during the working day.

The benefits divide into service benefits and productivity benefits. Service benefits include the ability to carry out most actions at branches in a fraction of the time which they took previously. Company personnel are now well-informed. Productivity benefits include a great saving of staff, 160 overall. The life departments staff halved with a saving of 90. The accounts departments staff was also halved with a saving of 40. The company's new business is now double their 1977 rates and staff numbers are 100 less. It was also felt that the system provided some staff amenity in that it gave a better working environment, more interesting, effective work, less tedious low level routine work, and improved job satisfaction.

The system was conceived in outline late in 1974. Development was planned to take place in major stages over a four-year period. The timetable was adhered to and the system was in place and working by the end of 1978. The financial benefits, adjusted for inflation, were greater than those postulated at the feasibility study stage.

The company is most proud of its instant policy. The system started with a quotations function. Working from a computer terminal, any of the staff at any of the offices can type in details of a proposed assurance, such as the age and sex of the applicant, the type of policy required, the intended sum assured, and see immediately on the screen a detailed statement of the cost and any other specific information about that particular type of policy. The enquirer can then change any of the details and get a further quotation, or request the system to quote for a different type of policy for the same person. If the quotation is wanted in writing, the clerk presses another button and a properly printed document is produced, with the appropriate text and clauses, and the specific detail is instantly amalgamated by the system. The implementation of this innovation was preceded by a comprehensive long-term strategic plan which a high level team representing users and technicians was given two months to produce. The plan was approved rapidly and the first development phases were quickly implemented.

When the quotation design was well advanced, work began on the second of the systems' functions, the issue of the new policy. The system takes a completed proposal form and produces the policy there and then at the point of entry. It puts the policy "on the books" and the whole transaction is done in real time. The policy is printed locally where it is required and the customer can have the policy within minutes. Once the policy is on the books, the system maintains it thereafter. It also has extensive facilities to quote surrender, loan, claim or maturity values, and systems exist within it to associate together the several policies held by any one customer.

Along with this particular innovation, the company gives continuous attention to the application of emerging technologies across the whole spectrum of business administration. Established within its computer organisation is a business systems division whose role is wide-ranging and extends into any area of the company. Business systems take over where specific major applications systems stop. One success of business systems was the expansion and development of micro-film facilities.

The aim now is to move towards electronic mail. The first steps were taken when a communications network was established in 1975 between head office and all the branches. In 1977 a message system was introduced, which although regarded as rudimentary was important. A message typed free-form into any visual display unit is dispatched to any other screen or printer or alternatively to all stations. Up to two full screens are allowed and transmission is effected by pressing a single key. There is no editing or checking, no hard copy, and no storage or retention. A study of the traffic using the message system led to the development of pre-written messages which the sender could call up from storage as required. These facilities were introduced in 1979. At present some 80 pre-formated messages are available. The company is also interested in word processing. Some small steps towards it and a planned upgradeof the conventional typing process were taken in 1975 with the installation of work stations spread over three functional areas. The installation inaugurated in 1979, was doubled in size within weeks.

The word processors were then given access to the company database held on the mainframe computer. Branch offices use IBM display writers. The company is moving towards clerical word processing, and aims to reduce or eliminate the word processing operator from the production chain, allowing the principle (i.e. the clerk, the underwriter, or the supervisor) to produce for himself the finished letters from previously stored text.

A major part of the company's business comes to it in the traditional way through insurance brokers. Having dispersed computing power to its branches, the company wishes to extend this to brokers, and provides a service of quotation enquiries and messages which could well be followed by a direct policy service. At first a simple teletype printer was used for the operation, which although effective, was slow. The system was radically improved with the British Telecom Gateway enhancement to its Prestel service, which has led to the development of a cheap and effective way of communicating with brokers and of providing them with information and service. Prestel Gateway allows messages from private computer installations to be accepted into and routed round the network of Prestel computers. Any computer linked to Prestel, and anybody with a T.V. set which can gain access to Prestel, can send messages one to another. Transmission is by telephone lines, at electronic speeds, and all the calls over whatever distance are charged at local call rates.

Brokers participating in the company scheme are equipped with Prestel T.V. and keyboards from which they can, through Gateway, obtain quotations for life assurance direct from the company's computer in the same way as Company A's branch offices. The equipment can be static and permanently installed in a broker's office or it can be portable. The company expects to develop the system by offering policy issuing facilities and other information services. The data processing manager of Company A had been essential in the implementation of these systems; he stressed that the company was now using information technology for increasing new business, rather than just for the more efficient completion of conventional tasks.

In describing innovation in the company, the last eight years were regarded as the relevant time period. Intensive technology was used, first for cost efficiency which was successful; business has trebled with a net reduction in head count. It was soon realised that service and marketing had a greater spin-off than expected. The biggest single reason for growth in the market-place and in sales has been the use of new technology. Company A is now using data processing innovations for marketing purposes, and for reaching the customers in a fast and spectacular way. The aim now is to increase revenue rather than efficiency.

Innovation has had three phases of growth: first, cost efficiency; second, service; and, third, marketing. The emphasis on the earlier aim remains but has been superseded by others. The early decision to innovate was not the result of competitive pressure, but arose from a wish to do things better, and to reduce the armies of clerks. Staff costs were going up and machinery costs down, and the company was tired of the frustrations associated with the old-fashioned system. The reaction to these problems was almost emotional

The decision on the innovation process was made rapidly, helped by Company A's change of chief executive and top personnel. The new life assurance man was rapidly promoted to Deputy Chief Executive and in 1974 the company recruited a new professional team for computing. It formed a "computing committee" which had Board of Directors approval for its efforts. In two months it had laid a plan. The whole process was "very flash and quick". In January 1975 the team got board approval for a four-year plan. The instant policy which it developed is still regarded as unique.

The factors which affected the speed of progress of innovation in Company A included the following five items.

1. Having the right man in a position of power to push the project through. The champion has to wish to do things, and to be fairly ruthless in getting them done.
2. The right innovation team—in this case the computer man and his team.
3. A partnership between factors 1 and 2 and this helps to speed up the process.
4. Stable industrial relations.
5. Adequate funding for the project. This is secondary since funding is rarely a problem if the other factors are present in the company.

The technologies essential to the innovation are all incorporated in information technology. Insurance had to change and methods for handling it have also had to change: "We need procedures and attitudes of mind that go with high technology. The management style must change and personnel policies must change". There were some problems associated with the speed of supply of items for innovation: for example, there were equipment problems with printers in the branches. Manufacturers' software was also criticised for its many shortcomings. The company writes all its own applications software. Technical delays should be regarded as temporary impediments rather than real delaying factors. The major constraint has been the time it takes to build systems, and this remains the major bottle-neck for the future. Company A has doubled its internal productivity on software, but it still takes "an alarming amount of time" to develop. There are limits to the growth and change that a company can take at any one time. Company A personnel have adapted very quickly and their adaptability has not proved to be as substantial a constraint as some anticipated.

Initially, the cost efficiency motives were pushing innovations, so strongly that market assessments did not enter into the equation. However, it soon became clear that new marketing possibilities had been opened by the success of the system. However, the company remains unsure of its marketing strength and tends to operate more on the basis of "entrepreneurial hunches". Company A makes a very detailed cost assessment. Analysis showed the usual reasons for escalating costs such as overrun, manpower, and machinery, were costed. Company A was initially uncertain about the accuracy of its manpower plan, but it proved very accurate. The machinery costings were inaccurate, but the company was saved from the consequences of miscalculation by the fall in prices of the machinery that it was buying. Thus the final expenditure was close to the initial estimates of cost.

Secrecy was not considered to be of vital importance. Insurance is, within limits, an open-door industry with an atmosphere of co-operation between financial and commercial companies. It was suggested that this trend is dying and people are now being rather more secretive, particularly in banking. Secrecy becomes important if the innovation is technically simple but still at the leading edge of competitive change.

The internal decision-making process involved in these innovations in Company A was at the very top level. The company decided not to follow the usual patterns, of a long-standing part-time committee making recommendations, and then having to endure a very long process to move their recommendations through the hierarchy. Instead, it implemented its changes in a short timescale through a small full-time committee of very senior people with links straight to the board and the chief executives. The committee was task orientated. This company commitment to data processing is still strong and the chief executive gives one morning a fortnight to computing matters.

The career background and training of the senior personnel involved in innovation decisions was suitably technical. The respondent felt that one needs one good technical thinker with technical imagination and then a meshing of appropriate disciplines. The risk of innovating was felt to be less in service industries where the technology is sound and the costing easier. To be innovative in the industry requires using outside innovation and established technology. The pioneering area is in applications rather than technology, and the risk is mainly social, in that customer reaction might be completely mis-read. This respondent stressed that a major champion for an innovation is essential and in the case described the champion had been the deputy chief executive. To talk of one major champion might mislead since in reality enough champions to form a "gravitational pull" are required. When Company A was innovating, there were strong opponents who expressed particular concern about the cost and workability of the system and who argued that the change was not necessary; in this case they did not prevail.

Company A has a technical division concerned with existing computing equipment,and a nucleus of people who think of the future, but it has no dedicated R&D area. The company view is that this type of technology works best if it is handled as part of daily life.

Company personnel profiles are now different. There has been loss through natural wastage and the company has closed its Scottish office, and has also had redundancies. Initially it had a hundred redundancies but since then there has been sufficient growth in business to mitigate further loss. Patterns of working have changed; there has been a breakdown of specialists and a move toward a "Jack-of-all-trades" approach. The need now is for-all purpose people, not specialists in a particular risk area. The intelligence and risk taking is to some extent invested in the computer. Jobs are broader, and slightly upgraded because the person using the computing equipment must solve all the problems without referral, so the job satisfaction level seems to be higher. The company expected an increase in qualifications at all levels of its personnel, but this has not been very noticeable; increases have been marginal and have probably coincided with a market availability of more highly qualified labour. At higher levels Company A is a very professional shop, and there are more highly educated people than there were. There are 30 programmers who are better skilled than those the company had previously. There has also been a substantial upgrading of personnel in the software area, which now has a 100% graduate intake. The company is now undertaking an upgrading process in the computer area.

Company A appears to have recognised the importance of the factors listed by Rothwell[49] as likely to be associated with successful innovation. The company weakness in marketing has been noted and an attempt made at improvement.

Company B

Company B is a British company, the interviewee was the Data Processing Manager, who places Company B, as probably the leading company in the UK, as the largest composite company on a premium income basis and as the leader in distributed processing, although another major company was running a close second. In the non-life area the company has a network linking the central computer to a number of smaller computers in branches where it holds 95% of the information on policies except the large commercial ones. Company B has a philosophy of duplicating files, one in the centre and one in the branches, and can therefore proceed with business in any branch for three of four days, even if the main computer is not working. A customer can telephone a local branch and a number of people can access the policy or claims record and, for example, change car insurance on the spot. Changes go through quickly at the close of business to the main computer so that the next day each branch has a new up-to-date file. Company B also does real time updating. The auditors dislike this because it may confuse audit trails. The company can search its files for vast ranges of information, which it regards as a very important facility. Company B is a big user of laser printers, and now has three, so that the quality of what is sent to the customers is greatly improved. Customers can pay in a variety of ways, and

49 Rothwell, "Policies and Industry", Op.cit.

inflation factors and all relevant details are on the renewal notice. Company B now provides marketing messages, and advice on renewal notices, all at an increased speed. In the event of an emergency (e.g. a flood), a team can settle claims extremely quickly. The architecture used is an IBM 8100. The company also uses IBM software, but it does use some other manufacturers' devices. It is concerned with office automation, distributed systems and word processors and is attempting to merge text and data. Company B has its own telephone system, which combines speech and data throughout the country; word processor messages can be sent to Singapore or electronic mail to somebody in the same building. The Company B respondent felt that many advances were technically possible at the moment which, although not yet cost effective, would be in time. He felt that the IBM personal computer has changed the market very considerably and that IBM's new small personal computer could have a further significant effect.

Company B's key to successful innovation in the data processing area was to pay attention to design. The overall strategy must be designed carefully because any wrong moves can be expensive for the company. The company is particularly application oriented, and has an investment system which it regards as a leading edge system. This is now being sold by software houses on its behalf. It also has a marine insurance system which it might sell in the future, but at the moment it provides a major advantage for the company. Insurance was suggested as particularly amenable to data processing innovation because of the amount of paper and messages being passed. Labour reduction was regarded as an on-going process, if companies are prepared to accept the consequences.

The decision to innovate was affected by the whole branch network. There was pressure from the general manager on data processing to improve the service to the insured, and the company was becoming increasingly bogged down in paperwork. Very great growth in business had to be accommodated. Although staff reduction was not the initial objective, increases in data processing led to increased flexibility in organisation and staffing. Before mechanisation, restructuring of any type was a major job; now it is easier. Increased use of data processing also shortens lines of communication. Senior management and technical personnel can access information easily and quickly and this increases their productivity. These trends could lead to reduced numbers of layers in the management structure. Middle management are most vulnerable to this type of change. Volume of business could now be doubled with no staff increases, if the business is there. Important mergers were seen as likely in the industry as were moves towards financial conglomerates similar to those in the US.

The initial impetus for innovation was to cope with the volume of paperwork and to ensure a better service; now the impetus is to provide new products, to reduce costs, and shed staff. The next possible innovation being discussed within the company and on which experiments are being conducted is to link brokers to the company. The Company B respondent felt that it was hard to estimate how long it might take to turn ideas into operating systems, but that it always takes years rather than months. Once the system is working however, enhancements can be very rapid. Many innovative ideas come from data processing.

Once a prototype has been developed most systems are evolved in cooperation with users and then are returned to data processing to be made suitable for daily use. The increase in user involvement has made the cost effectiveness of systems easier to estimate. Speed of progress can be affected by the skilled resources available. Good people are needed to design systems and produce software, and as there is still growth in the data processing market, recruiting personnel is generally rather difficult. New businesses are growing up all the time and these tend to take people from those large companies who train their own personnel. Equipment on the whole is less problematic, although software often requires a substantial wait until suitable software becomes available.

In Company B much innovation has been dependent on telecommunications which, although much improved, were felt to require further improvements at a better price before more innovation will be possible. There is a great need for developments in fibre optics and increased use of satellites. The slowness of British Telecoms was regarded as a major constraint and hopes were expressed that the freeing up of the monopoly would help in this respect. British companies were thought to suffer disproportionately since flexibility within the UK telecommunication industry did not match that in North America.

In Company B "connectivity" (an IBM word) was regarded as important. The need for safety required a major supplier, since the costs of change are enormous and can run into millions of pounds.

IBM is regarded as somewhat conservative. In most areas it is possible to find a better individual product than IBM can supply, but when it is all put together, the IBM system was felt to be quite unbeatable. Company B buys disc drives, tape units and screens from other manufacturers, but ensures that they are all IBM compatible. At present a great deal of money is being spent on these items. IBM is seen as a very market responsive company with a clear awareness of market pressures. It was suggested that this responsiveness was another of the major reasons why the industry went with IBM despite its size and general conservatism.

Problems with software in Company B are common. Obtaining software to specification when it is new takes time, but such problems must be anticipated. They can slow down innovation but are not a major hindrance. Hardware is rarely problematic if its purchase is planned well. Where demand on manufacturers outstrips their ability to supply, as has happened with the IBM personal computer, it can cause difficulties, but if plans are made far enough ahead of the required delivery date, this should be overcome. Company B's biggest difficulties still lie with British Telecom.

Company B allows a great deal of latitude in decision making on innovation. Decisions are made by the UK executive users of management information services (data processing) who present a joint proposal to the executive. The head of management information services sits on the executive. A proposal would be considered in line with the existing work load, and would be given priority. The company at the moment has more innovation work than it can possibly handle and so attempts to push the best schemes to the front. Many early mistakes resulted from lack of consultation with users and from presenting users with too firm a product. About two ideas per year are explored by management information services. The heads of the four management information services in the UK, USA, Canada, and the Netherlands meet to discuss progress, to explore innovations in the company, and to consider the transferability of successful schemes. The demands on the time of management information services have been very high. Problems have arisen in satisfying needs rather than in selling management information services to the rest of the company. Often when a problem is being solved, a more sophisticated system can be sold to the user.

The momentum for change at the moment in Company B is enormous, and the company is thinking in terms of cost effectiveness rather than cost. Cost benefit analyses are carried out on major projects, but it is often difficult for users to express benefits in tangible terms. Company B takes an interlocking approach involving the users and the management information services in expressing likely costs and benefits. The view of the finance group is then sought. As a result project control and estimating has improved greatly within the company, and the management information service department is now good at expressing costs. The user still has a greater problem in expressing benefits with equal clarity, and it is important to ensure that costs are not over-emphasised, and benefits under-emphasised. Considerable help is given to users to express benefits in a way acceptable to senior management, to transfer the seemingly intangible to the tangible, but there is a grey area in the middle which is harder to define. One method to help users express probabilities and benefits has been to make them think in ranges. The company uses an investment appraisal approach and monitors the project to see how it has worked, and to access the benefits firmly. This process also helps in recognising unforseen enhancements: for example, one development not initially anticipated was the ability to index-link policies.

The Company B respondent like the Company A respondent regarded secrecy as of little importance within the industry. Secrecy was felt to be too strong a word. The industry, and particularly the data processing part of the industry, has always been open since the beginning. Systems generally were shown around, although this is now slightly less likely. There is an attempt to maintain a leading edge for a short time, perhaps for two years at the very most. Implementation is the real problem.

The backgrounds of the personnel involved in innovation in Company B are very mixed. Some have a predominantly data processing background, and some an insurance background. Those from outside must gain insurance knowledge, but unless one is an actuary or an underwriter, the type of insurance knowledge required is not overly complicated. For Company B the keen computer person is seen as the most likely to be innovative. The insurance people will then become interested in the possible benefits from the innovation.

Again an innovation champion was considered important, but the champion needed to be working with top management who were mature enough to support his ideas and to let them run their course. The champion must be strong enough to foster programmes if they appear to be useful, but to kill them when they no longer seem likely to succeed. This ability to stop programmes was seen as significant as the ability to start them. Since the insurance business is very competitive, the need to innovate is vital, but insurance is rightly also a cautious business. All the companies produce very similar products and the scope for innovation therefore is limited. The emphasis is on what one can give, how one presents it, and how efficiently one can organise it. Data processing has a large role in all of these activities.

Companies A and B, which both claim to be and can be regarded as innovative within the insurance industry, provide an interesting contrast with the Canadian and American companies involved in the project who made no claims to being innovative in their use of data processing, nor did they desire to be regarded as innovative in this area.

Company C

Company C was founded in Canada in 1900 and offers a full range of life insurance, health insurance and pension programmes, on a participating and non-participating basis to individuals and groups. In addition it provides re-insurance services to other insurance companies. It serves nearly 5,000,000 clients through 260 sales and service offices in Canada, the USA, the UK, the Atlantic and Caribbean Islands and Hong Kong. Company C has assets of $3.4 billions (Canadian), and more than $48 billions of life insurance in force. It ranks in the top 2% of life insurance companies in North America.

Company C has as a subsidiary a major computer service firm serving the North American market. A data processing subsidiary was formed in 1972. The company also has in-house data processing and all of its software and applications are in-house. Company C represents 10% of the data processing subsidiaries' revenues and has shared processing and is diversifying. The technological subsidiary is expanding into high tech fields of fibre optics and software.

Company C's major innovations at the time of the first interview concerned the development of an information centre and desk-top computing. A leading edge position was regarded as too risky, too expensive, and to have no particular advantages, Company C wished to be a solid follower. Company C still regarded itself as comparable with other leading companies in Canada, definitely better than its major competitors in some areas, and as good as any in most.

In banking it was suggested systems advances meant competitive advantages, but this relationship was not so clear in insurance. However, new products need efficient data processing and the development of new products, and the data processing requirements for them, were integrated in Company C. Innovation in the company in the last few years had been more concerned with marketing support than data processing.

The information centre development in the company is concerned to encourage end users to use data processing directly. Actuary managers, clerks and marketing personnel should have access to their own data processing facilities. The information centre also has a training function and has 40 microcomputers throughout the company which are heavily used. The scheme is regarded as an internal company marketing project, and the number of microcomputers is going to be increased in the next year. Company C is also establishing a network for integrated electronic mail, and moving toward an electronic office system using desktop computers. The time allowed for the development of this last project is between two and two-and-a-half years. It was suggested that other companies that have used micros as a sales marketing tool have paid the price of being the leaders in the field, and have often been left with a technology which is no longer the most efficient. The information centre development was seen as reflecting an industry and data processing trend to push responsibility onto users. Previously such moves had been inhibited by a lack of suitable tools. These are now available. Company C, like the other insurance companies in the study, rely heavily on IBM products.

It was felt that all innovations within Company C, and probably everywhere, took too long. An innovation champion was seen as very important. The champion needs to convince a sufficient number of people to support the innovation effort, and the speed of the development may depend on the rank and importance of the champion. In some companies it was suggested it does not pay to innovate because the corporate culture is inhibitive. If a company is conservative, a situation arises in which there are few failures, because there are few attempts at innovation. Failure, therefore, rebounds badly on whoever is associated with it. This leads to a company in which there is no incentive to prove oneself innovative, and in which one can live in a steady state. The survival of this type of organisation is aided by a system of promotion from within. Company C, it was suggested, needed more promotion from outside of people with a data processing background. Company C was at this time beginning to change slowly, from a company with a conservative corporate culture, to one in which innovation was seen as having significance.

The new technology necessary for innovations in Company C has come from IBM. They have been studied within IBM and then bought by Company C. Company C is using fourth generation computer languages which are still being developed and are designed for end user use. It is expected to take about nine months for these to gain general acceptance. Personal computer spread-sheet applications are of prime interest, but there is concern with the increased error risk with personal computers which was felt to be higher than under the old clerical system. The company is working on ways of overcoming this problem. Concern was also expressed about the security of the database and the lack of audit trials, problems also mentioned by Company B. For Company C "more users mean more risk and more errors", but for Company C, advantages of increased data processing can be found in the more rapid access to data, and the speed of report retrieval which is a major function. One-off jobs become practical as does the use of sophisticated graphics.

Innovation within Company C has not been held up by lack of funds. It has been slowed by its suppliers, particularly in software. The company has not yet experienced problems recruiting personnel at higher levels and has recruited mainly from within.

In Company C, there can be a high degree of latitude in planning a new system, providing an innovation champion has gained support for the change. At the time of the first visit a major need in the company was felt to be pushing managers on rather than reigning them in. Innovations, it was suggested, were often stopped at lower levels by an attitude of " . . . get on with your present work and don't think about the furture". Many possibly interesting innovations never received realistic appraisal. This contrasts markedly with the attitudes expressed in companies A and B and with the situation which has now developed in Company C.

Marketing problems in data processing have not as yet arisen in Company C since data processing related product developments are rare. Generally users would be asked how far data processing, was related to their products after the product had been developed. No new products have come from data processing whose role has been to influence but not to develop anything original. Company C's UK subsidiary has a joint development system, with data processing working in association with marketing to develop the perfect product. The UK and Canadian situations were seen as radically different with different tax systems and different markets which require a different marketing approach. Cross fertilisation of ideas was therefore considered to be irrelevant.

Innovations in Company C were not seen as overly cost sensitive although developments were costed closely. For the innovations mentioned here, development cost was small relative to total budget, and often hard to justify in cost-benefit terms. The ideas had had to be sold rather loosely. Getting funding for any corporate activity within the company was not easy. Within Company C the internal decision-making involved in any innovation tended to be informal and depended on where the original idea came from and the level of funding required.

The career background and training of senior personnel involved in innovation decisions in Company C tends to be a mixture of data processing and insurance skills.

Company C's attitude to the importance of secrecy in developing innovations mirrored those expressed in Companies A and B. The factors considered significant in being an innovative company were the availiability of funds for a project, and a backdrop of successful projects which allows people to feel that interest in innovation, change, and in risk taking is worthwhile. A corporate climate which encourages innovation and which must come from the top to provide the appropriate leadership was considered of great importance.

Initially Company C seemed in many ways to be a company in change, attempting to move from a fairly conservative approach to a rather more innovative one. The respondents felt that commitment to innovation was probably still not strong enough throughout the company, though it was now stronger from the top. The recognition of innovation as a corporate task was also rather weak, and therefore any innovator was unlikely to get sufficient support, however well designed his system, and however productive and innovative it might eventually be. The task facing an innovator in getting the company to accept a new idea remained fairly formidable.

A further visit to Company C showed that the company was changing faster than had appeared at first. Company C has a new strategic plan for the insurance company and has gone from spending very little on data processing to an $8 millions expenditure on data processing development. Company C now feels that its work is quite innovative in the insurance industry. The subsidiary, company which was an alternative to in-house computing, is relatively profitable, and has a new president. Company C has bought five subsidiaries including a computer corporation and a fibre optics company. Changes are now very rapid within Company C. The company is now benefitting from a "snowball effect, so everything is moving faster than was anticipated".

The company was at one time a conservative group with available resources but uncertain as to how to use such money. This is not a problem the company now experiences. Company C has set up a company to escape insurance regulations, and has bought a trust company which it is going to expand and which will be a companion to the insurance company. One illustration of this change in Company C's philosopy is that until recently Company C has never advertised, but this year it sponsored a major sporting event. Company C has also recruited new people with non-insurance backgrounds who have not been used to the conservative environment of the financial world. There was hesitation initially about this move, but now the company feels it should have done it earlier.

The organisational structure of Company C has been flattened. It used to be very hierarchical with about 17 layers between the office boy and the chief executive; this has now been cut down to seven layers, and the aim is four or five. Underlying the change is the notion that ideas can now get through the system much more easily. The respondent suggested that there has been a change in North American thinking about the number of people any one person can manage; the idea of what is possible has been expanded and this has enabled Company C to eliminate a whole layer of management.

The chief executive of Company C now talks with people in the field, through agents' advisory committees. This was partly an agents' idea, and resulted from a meeting at a convention. The group has now met and seems likely to succeed since it means that middle management can no longer block innovative ideas coming from the agents themselves. Company C is interested in "participatory management" and wants to improve communication.

Company C is now concerned with market research and has a market research company. In the past, Company C has never spent funds in this area and used to rely on the joint insurance group.

Company C's insurance management, it was suggested, used to be too narrow. The company has now set up links with universities and runs short, concentrated courses for personnel who are being exposed to new ideas and hence things are changing fast in the company. The aim is for change to become the norm and not the exception. Company C is learning to manage change and to cope with it. Company C has now introduced ideas of profitability by product rather than by relying on the old actuarial basis of business analysis. Another Canadian company is also moving along this line, but it is not common within the insurance industry. Company C now feels that "the ball is rolling nicely and we do not want it to go too fast or too slow". At one point the company went from being too conservative to being in danger of changing too fast. It reduced staff by 25% and changed many other things. Company C was on the edge of tolerance of change. Now the company is getting better financial results and so it is beginning to see the rewards for what has been endured. The company now plans to buy a new insurance company in the US and to sell insurance to a lower income group and also plans to go into banking and attract savings, an area in which it feels it has lost a good deal.

Company D

Company D, also Canadian, rejected the label of innovative but aimed to ensure that it was a leader in its use of information technology, and was committed to information technology as a primary vehicle to deliver services to the consumer. Of particular interest was the interface with the consumer, but there is little innovation here as yet. Company D, unlike Company C, felt it was a leader in its use of technology. The respondent, the data processing manager, had been with the company since 1979. He came originally from manufacturing, not from insurance, and suggested that there is greater leverage by the application of information technology in financial services than there is in manufacturing and distribution, " . . . insurance is in some ways an information processing factory".

Company D was founded in 1874 and today operates across Canada providing life insurance and financial services to more than 2,000,000 Canadians. The fourth largest life insurance company in the country, it is a leading investor in mortgages, real estate and the financial markets, supplying funds for a broad range of Canadian businesses and developments. It operates a national network of regional offices with a sales force of more than 2,200, supported by 800 administrative staff and a corporate headquarters group of 1,700 employees. It employs the latest in electronic technology and its overall objective is to be the leading Canadian life insurance company. Despite the effect of recession, the company continued to grow in 1982 and paid out $151 millions (Canadian) in dividends to policy owners. Earnings per share increased by 3.2%. The company assigns a high priority and considerable resources to information technology systems because it sees the aggressive pursuit and application of leading edge office systems technology as vital to productivity improvement in competitive effectiveness.

In the late 1950s and early '60s, Company D had a leading edge in DP applications. In the late '60s and early '70s it was interested in hardware and software and was a data test site for IBM. In 1974 it was the first user in Canada of software database coast-to-coast, for its 130 offices, in this respect it was the leader in North America. The company has reduced its administrative costs, and its numbers of clerical and record keeping people, and these activities have been cut substantially within the company. The early focus was on clerical productivity; there was little focus on management and professional staff. This was fairly typical in the industry and does deliver significant improvements.

For Company D the early savings from the introduction of data processing were easy to see and easy to deliver. The second stage which involves the management and professional staff and new business is more difficult. The third thrust of the programme is to continue to improve administrative productivity, to begin to find ways to support management and professional staff, and to begin to find ways to integrate technology with the sales process. In Company D the key factor in the decision to innovate is the end-user, then a high level executive sponsor, budget, and outstanding creative and innovative people are needed.

The data processing manager's responsibilities fall into the six areas listed here.

1. Business systems planning.
2. A systems development group which is sub-divided into four, reflecting the lines of business in the company.
3. The development support group which focuses on data, not on applications, and is concerned with the productivity and quality of systems.
4. A technical staff group concerned with hardware, software, telecoms, vendor relationships, etc.
5. Computer services which are concerned with day-to-day operations with the word processor complex, and the 900 terminals.
6. An office support system which is concerned with the information centre and with helping users to help themselves.

Company D has an annual corporate planning process which involves each function. The aim of data processing is to penetrate the organisation with technology and to make users self-sufficient. End-user computer languages are wanted to help this to happen. So far 350 non-data processing end-users have been trained. The company's office systems technology is moving towards office automation and the electronic office.

Innovations in Company D involve two types of people. First are the technical people who are highly skilled and decide what is appropriate for the company, whatever the problem, simple or complex; they are in charge of its technical side. If the idea is adopted, then the second type of people become involved. They are skilled in inter-personal relations and can sell the system to the end-user. Decisions about innovation in Company D take a variable amount of time depending on the nature of the project. At the moment, the company is engaged in projects to analyse the impact on sales of providing micro computers to its sales reps. The implications are both economic and social. A 15 to 18 month period has been allowed to sell the pilot study, four months to plan it, and nine months to gain experience and to judge the likely increase of productivity of sales reps.

The head count in Company D is now flat or going down slighty, but productivity has gone up considerably as has the wage bill. The structure and employment profile in the company is projected as changing from a traditional pyramid to an oval shape with accompanying lengthening in the spans of control. The company is in a process of evolution rather than revolution. Work may expand to fill the technology and time available " . . . you must be careful to get more value rather than more capacity, we are interested in making money rather than saving money".

Factors which affect the speed of progress once an innovation has been adopted are organisational and concern lead times which have been reduced and software quality which is rising. Social and management problems are possible but technical problems nowadays are rare.

The quality and supply of software and the end-user interface are still problematic. Company D is very concerned with improving user-friendliness which still has not reached a desirable level. The technologies essential to Company D's innovations derive from the electronics industry. Hardware although never perfect, is now very good. Deficiencies in software, however, are considerable. Company D has very few problems in attracting suitable personnel, it is in a good location and has a low turnover of staff.

Company D is always concerned with profitability and does cost-benefit analyses of all new projects. All decisions are business decisions based on careful market and cost assessment. A business case is always developed, costing is very close and there is an attempt to measure benefits specifically, although this is difficult. Company D agrees with the other companies in the study about the nature and importance of secrecy, but the respondent found this attitude particulary surprising coming as he does from a manufacturing background where secrecy was of much greater significance.

The internal decision-making process concerned with innovation within Company D varies. Typically, there is a steering committee and a committee concerned with the allocation of information service resources, which meets monthly and is made up of the eight vice-presidents plus the respondent. The company has a five-year planning exercise, which goes forward to the president. The plans are then submitted to the company's budgeting process and are allocated a specific budget, a headcount and a plan. Approval to spend is then sought.

Most of the senior personnel involved in innovation decisions have considerable insurance experience, but their initial backgrounds are extremely variable. Their data processing backgrounds might well have been in a non-insurance area. There are many people within the company with a pure maths background and Company D favours Co-op courses, which are similar to sandwich courses in the UK. Company D annually recruits four to six graduates into information services. The company has attempted recently to get an influx of new blood. The current president was a consultant recruited in 1978 to analyse the strengths and weaknesses of the company, and the following year he was made vice-president. He favours a policy of integrating into the company executives from outside the company and from outside insurance. Two of the four vice-presidents have come from outside insurance. The aim of this integration exercise was to bring a different perspective to the company.

The importance of a major champion for an innovation in Company D was stressed and this champion needs to be an executive sponsor. Although the data processing group can take initiatives, success requires a sponsor. At the moment the company president is very much in favour of data processing initiatives and all of the eight vice-presidents have at some time sponsored particular projects.

It was suggested that the changing nature of financial markets means that Company D needs to be more aggressive. The respondent felt that Canada was over populated with banks, trusts etc., and so innovation was very necessary to be competitive. Company D is concerned with the aggressive use of technology which is mandated by the competition merely to stay in the race. It is concerned also with personnel, with training, re-training, and the quality of working life, which were seen as important in their type of company.

Company E

Company E is mutual life insurance company which has been established for 140 years in the US. The company operations have recently been reorganised by product lines. This was to allow a sharper focus on the public's growing financial needs and provide the resources necessary to service this growing market. Four regional vice-presidents have been appointed. During 1982 the company's total assets grew by almost 10% to reach $8 billions (US). The assets under management by the company at subsidiaries and affiliates grew to $18.3 billions at the year end, a 24.5% increase over 1981. The goal of Company E is to be integrated, comprehensive financial services organisation. The company sees itself as market driven and growth oriented.

Company E is concerned with the application of technology to product development. The recent history of insurance has witnessed the move from rate books to the computer era with an interactive terminal for unique, specific policies. Company E can now match the ultimate product to the market niche more closely. The company's computer sophistication is tailored to individual needs and to new products. It does not for example have the old cumbersome system of add-on policies; instead it has one policy per life which is adjusted mathematically and automatically by computation. The developments for this facility took 25 man-years in maths and software.

For the Company E respondent innovation in the insurance industry lies in the ability to apply technology to updates and exchanges on a continuing basis. All of the company's customers had had an exchange offer, which gave old customers access to new computerized policies and involved a massive data processing project. Company E was the second US company to offer this. The company also has a financial service and planning division.

Insurance and finance present many data processing problems, large and small, and there is a major organizational effort to solve them. Company E has invested 15 man-years in its financial planning system. It is replacing 70% of its 2,000 terminals with micro-processor based terminals and using IBM display writers and personal computers. Its customers have an average income of $50,000. In Company D $40,000 income was regarded as high-income bracket. The financial services offered by Company E are developing out of the life insurance so the company is not now just providing life insurance. The company uses brokers all of whom are dedicated to Company E. The brokers, not the policy holders, are regarded as the clients. It needs to attract the best brokers possible. Company E is interested in technology driven marketing strategies and innovative products using computer support. Computing changes constantly increase capacity with reductions in costs and size. Even IBM, the respondent pointed out, is now attempting to reduce its costs. This echoes the view of IBM as conservative but market responsive put forward by Company B.

Agents need their computers to be user-friendly but technically sophisticated pieces of equipment. Equipment incompatibility remains a problem, and therefore intelligent vendor selection is necessary. For the brokers, although hardware may change, software and their knowledge of the insurance industry will last, if they pick the right insurance company, in this case obviously Company E.

The primary motive for the introduction of data processing in Company E was the need to increase cost effectiveness. Now thinking is market driven, and cost is no longer the primary emphasis. There is, of course, an element of survival, as was mentioned in Company D, "One has to do this now in the industry, but you don't just need money to do it, you need talent and the latter is harder to find and more significant".

Decisions about innovation in Company E are very fast because of the organizational structure and because of our respondent's position and background. He is an executive vice-president, a data processing man turned user, who has a quarter of a million US dollars in his R&D budget which does not require justification through the normal processes. There is, of course, always access to more funding if it is needed, but this must be justified in the normal way.

The speed of progress of innovation is crucially affected by the major supplier, IBM. The insurance business, not the technology business, is Company E's major concern so it has chosen an IBM vendor. This approach is typical of the insurance companies in this study and also further confirmed by Barras and Swann[50]. As the Company E respondent commented, "As technology merges you need outside skills to lash it together, if you do not choose one particular vendor. Of course, there are risks both ways round, either you become too dependent on one company, in this case IBM, or you spend a lot of time and effort trying to lash incompatibles together and conversions are very slow and very expensive". By and large, IBM has the whole insurance market in North America. Company E is waiting for a breakthrough in direct access storage devices and in software for index retrieval, and feels IBM is most likely to solve these problems and that ultimately all insurance companies will depend on IBM.

---

50 Barras and Swann, Op.cit.

Company E regards its policy of buying from one vendor as the only possible solution in the long-term. In the short-term it recognises more advanced items are available from other companies but the various costs involved in the initial purchase of equipment, developing software, and running costs are too high to allow for a fourth cost of conversion to a new system. This strategy requires waiting for IBM to develop items which other companies already have, but Company E, unlike Company B, is prepared to wait.

There is little latitude in the design of innovation systems since they come from IBM. This is a major constraint, but within this there is a great deal of latitude on how to use these systems. Company E has signed a non-disclosure document with IBM and has worked very closely in the development of devices to suit specific company purposes. The reliance on IBM by major companies means that there are opportunities for small companies in the insurance business prepared to get the most up-to-date equipment available and risk later problems of incompatibility.

For Company E, market assessment is of crucial significance. There is an actuarial data processing marketing group which must interact with the sales force to outline marketing strategy. "If the sales force don't want it, they don't need it, and the company is not going to spend any time on it". For Company E innovation implies vision and providing solutions to problems. What is needed is an accurate perception of the world. "It is too easy to feel that your perception of the world is the right one and to invest too heavily in things which are of little use to the brokers." The company has a field agency system and training team who travel and talk to agents to find out what is needed. New products and processes have been developed in this way. Company E is in the process of acquiring a bank. which it feels will allow it to do things that nobody else in the US insurance market can do, and will create an image that will buy it talent. Thus it is prepared on occasion to take a risk and try to lead the sales force, but the usual policy is to wait until demand for a project is high before development goes through. "The search is for solutions and not problems."

In Company E, which is concerned with the potential return on investment, cost assessment is important and must be identified very clearly. At the moment the company has $1 million to buy software, which it regards "as throwaway money"—that is, sales support for financial planning.

Here the company is in an area of management judgement and aims to buy into a market; therefore it is difficult to calculate return on investment and Company E does not try. Another area in which spending is difficult to decide is that concerned with the retention of agents.

For Company E any advantages derived from secrecy about use of technology were of less significance that the need to be an open company to sell to agents. Secrecy cannot be a major concern, since innovation takes place in the context of the technical management team with a given perspective, which is not alike in any two companies. Even if another company knew Company E's strategy, it would be unlikely to find that it fitted with its own particular philosophy, and therefore would be unlikely to copy it. In some senses technology "concretes you into a system" and means that secrecy is of less importance unless someone has exactly the same system.

The internal decision-making process involved in the development of an innovation in Company E is related to the company structure. A discussion with the president and the chairman might be required but, for example, in the area of office automation few people are competent to judge, so data processing gives the lead to the company and is generally followed. Complaints arise only about short-term solutions which tend to result from short-term problem solving without a long-term strategic view.

The career background and training of the senior personnel involved in innovation decisions in Company E are much more strongly insurance oriented than in the other companies visited, particularly Company D. Data processing people have an actuarial background rather than a primarily data processing one; this, it was felt, aids communications with the problem-oriented user and takes solutions to the user. An insurance background was regarded as absolutely essential. Project appraisals in Company E are concerned with user satisfaction, keeping within a budget and being on time. The user comes first, the data processors second. The aim is solving business problems not DP problems. The need is to be knowledgeable about user areas, indeed some of the DP people in Company E have been recruited out of it by users. The Company E respondent's view is that the insurance area is more complex and harder to master than data processing, a view not shared in any other company visited.

For Company E an essential for innovation is leadership that is willing to take risks and put itself out on a limb. This suggests a good track record and, therefore, in a sense an innovation champion. Of key importance is an ability to deliver which creates a supportive corporate climate and makes the system work.

In the data processing units in Company E all necessary facilities are available. Within systems support there are industrial engineers, work flow systems analysis and work flow design. The company also has testing services. A purchasing officer for equipment in DP ensures that no non-IBM equipment is purchased. The overall company philosophy is to solve business problems. Company E is aggressive and wants to be a leader. Business problems are solved by the application of technology.

Although the Company E respondent recognised the importance of general economic conditions in allowing a company to be innovative, insurance was seen as riding through most economic conditions, volatile interest rates and inflation. The general financial position of Company E is extremely sound, but cash flow can sometimes inhibit innovation. Indeed financial position is so strong that funding innovation can never be a problem unless there is a series of catastrophes. The actions of competitors affect the company's world view and it will copy any ideas that seem good. The respondent felt that the role of financial considerations in innovation might be exaggerated: "There are leaders, watchers and followers, these are different roles entailing different risks, and with different results associated with each strategy". Corporate culture rather than the availabily of finance would be more likely to dictate a move from one to another role.

Company E does its own training and its staff turnover is low. It recruits people with potential and then teaches them all that is necessary. It has no problem in attracting suitably qualified personnel. Company personnel are now used to working in a dynamic environment where reorganisations are frequent. This is accepted because the company guarantees security for employees. The respondent explained: "You must have the right people, have them organised and working in the right climate. We are an insurance marketing company not a computer company. Even in data processing we are solving business problems and not playing with technology and this must always be remembered".

II. Materials and Resources Industries

THE PAPER INDUSTRY

Overview

The main constituent of paper and board is cellulose fibre. For economic reasons wood is the most widely used source, although straw, sugar cane (bagasse), bamboo, reeds, hemp and certain grasses can also be used. The use to which any fibre can be put depends on a combination of factors—availability, costs, tensile strength, opacity and bleachability. Softwood or coniferous trees produce the long, strong fibres used for both brown and white packaging papers, and the hardwood or deciduous trees produce shorter, weaker but stiffer and more opaque fibres which are blended with softwood to make printing and writing papers. These fibres can be extracted from timber by three main methods: using chemicals to dissolve the matter binding the fibres, by grinding the wood to release the fibres, and by partly cooking the wood and then dispersing the fibres by mechanical means.

The paper industry in concentrated in the US, Canada, Scandinavia, and South America. In Scandinavia they are now having problems in producing enough of their own trees, and are

Table 3

WORLD PULP, PAPER AND BOARD INDUSTRY CAPACITY, PRODUCTION AND CONSUMPTION 1981-1982 IN SELECTED AREAS

('000 tonnes)

	CAPACITY		CONSUMPTION		PRODUCTION	
	Paper and Board	*Pulp*	*Paper and Board*	*Pulp*	*Paper and Board*	*Pulp*
UK	3,844	192	6,680	1,493	3,124	141
Scandinavia	15,740	20,175	3,313	12,007	13,127	15,942
Canada	15,269	23,165	4,310	11,264	12,406	16,988
USA	63,683	50,966	57,966	45,024	54,133	44,786
Latin America ..	10,181	7,347	9,021	4,470	7,582	5,162

Source: Paper and Packaging Industry Annual Review, July 1983.

having to import pulp from the USSR. The influence of conservationists has been marked.

Most of the world's pulp is made in integrated plants, that is making paper as well, and only 10% is market pulp for use in non-integrated mills. Table 3 (page 81) shows capacity consumption and production of paper, board and pulp in 1981-1982 in selected countries.

The international paper business can be most easily understood in terms of four separate, but inter-related business areas. First is pulp bale form. In the UK waste paper is used as a raw material. Company H in the BNAC study is the UK company most heavily involved here. Imported wood pulp represents a large cost to UK paper companies and any ability to increase use of waste paper could have significant effects. Company F which is locating a new mill in the UK has been working on innovation in this area. Virgin fibre is still, however, necessary in many high grades and speciality grades of paper to meet stringent specifications of strength, burst, durability, and brightness.

The second business area involves substrate materials which working from pulp are turned into reels of flat paper, board or sometimes sheets. The third business area is concerned with the intermediate converting process, slitting, some printing, reeling, laminating, etc. The fourth business area involves a multiplicity of finished product including books, magazines, fancy packaging, silver paper, tissues, wall paper, etc. Companies can be involved in any or all of these activities.

There are, of course, other raw material opportunities open to the paper industry and these are areas in which experts feel that there is potential for innovation. They include the further development of pulp from grasses and from sugar cane. There is also room for innovation in the UK, for example in the use of straw and mineral fillers from domestic suppliers. Possibilities for the future seem likely in the developing area of bio-technology.

The paper industry is a continuous process industry and highly capital intensive. A new paper-making, or board-making mill costs up to £200 millions. To obtain an adequate return on investment, the equipment has to be fully utilised and the industry operates mainly on a 24 hours continuous working basis. All these factors act as constraints on innovation activity[51]. The industry is sensitive to

51 See J. Schmookler, "Economic Sources of Inventive Activity", *Journal of Economic History*, March, 1962; and Schott, 1981, Op.cit.

fluctuations in demand, and its fortunes are closely related to national growth. This explains the cyclical nature of the industry's operations. If economic growth falls off, so does the demand for paper and board and the profitability of the industry decreases. In these circumstances it is very difficult to keep capacity utilisation at a level which permits retention of market share on a world-wide basis. It is difficult also to provide the necessary capital investment to remain competitive. Innovation at the low added value end of the industry tends to be in processes rather than products and has an overriding aim of reducing costs. Industry sensitivity to GNP, to exchange rates and to energy costs all makes for a hostile environment for innovation. At the high added value end of the industry competitive pressures are also very great, and since product innovation tends to be neither fast, nor radical, even successfully innovative companies cannot gain an unassailable lead.

Innovations in paper-making

The basic technology (i.e. the Foundrinier paper machine depositing pulp into a wire and drying to produce paper) has remained unchanged since the middle of the last century. The industry has been, and still is, characterized by incremental innovations at each stage of the paper-making process. These innovations have principally been aimed at improving the speed of the paper-making process whilst reducing the level of "broke"—output unusable for various reasons. In recent years there has been the inevitable introduction of microprocessor based control systems so that in state-of-the-art systems the quality of the web is measured at various points and the results are analysed and used to control the flow from the headbox. Process innovations have also been concerned with energy saving.

There have been product innovations, for example, tracing paper about the time of World War II and since then there have been major developments in tissue making. The most recent product development has been non-wovens. These are not paper (i.e. cellulose) products but textiles produced on a paper machine. Non-wovens have found a host of uses in medical products, disposable clothing, etc.

Coating and converting

In many papers the paper-making process is only the beginning. Before being used they may be coated, impregnated, laminated, printed, etc.

Perhaps the major development in coating has been carbonless copying paper. This involves two coatings, one of which is encapsulated. When a typewriter or printer strikes the paper the capsules are ruptured and a chemical reaction takes place, to produce the desired image.

Other interesting coating developments have been dielectric coatings (which produce an image when a current flows through the paper) and thermal coatings (which produce an image where heat is applied to the paper).

There has been a tremendous number of incremental innovations in coating, laminating, co-extruding and impregnating papers for packaging. An interesting example is vacuum metallising, which involves depositing aluminium in a vacuum on to a paper substrate rather than the more common (and expensive) process of laminating aluminium foil and paper.

The future

Obviously this has to be speculative but the following are likely and may provide a framework for looking at the R&D programmes of papermakers.

1. Printing papers will tend towards the use of coated groundwoods (often known as lightweight coated—LWCs).
2. Office and computer papers will tend to be uncoated 'Xerox type' paper as xerography becomes a major printing—as well as copying—technology.
3. There will be continuing developments in specialist, low volume, high value-added paper or paper/other material combinations for use in packaging.
4. For board, as opposed to paper, a fundamental breakthrough would be the development of dry forming, i.e. paper from a dry stock rather than pulp, which would provide massive energy savings (energy at present accounting for around 10% of total cost).

Company F

Company F is an eastern Canadian forest products and packaging organization that ranks among the 30 largest publicy owned manufacturing companies in Canada. In the pulp and paper business area the most important product line is newsprint. During 1982 the Group's seven primary mills produced 815,000 tonnes of newsprint, mostly for export, and 500,000 tonnes of kraft pulp, paperboard and kraft paper. The company's capital expenditure for 1982 amounted to $242 millions (Canadian). The major expense was a project involving the construction of a chemi-thermo-mechanical pulp plant in New Brunswick to supply flash-dried CTMP to a completely rebuilt newsprint mill in the UK. Like the other companies in this sector, Company F has been affected by the recession and the consequent reduction in demand for its products. A drop in sales volume, together with continuing cost escalations led to a sharp decrease in operating profits.

Company F produces 1.4 million tonnes of paper and board; the largest US producer produces five million tonnes, and the largest in Scandinavia just over one million tonnes in several mills.

Adverse economic conditions, and a difficult financial situation in 1971, resulted in a major reshuffle in R & D organization in Company F, and reductions in personnel. From that time only a nucleus "back-bone" of researchers with "down-to-earth" programmes was left to serve as a springboard for future expansion. Thus, the present recession did not affect R & D significantly, at least up to 1982. On-going activities were maintained since these were mainly oriented towards manufacturing cost reduction, improvement of product quality, and mill operations. A reduction in total expenditure of up to 15% is aimed for in 1983 to meet company guidelines.

Company F regards its R & D outputs, both from cost effective and innovation standpoints, as favourably comparable with those of its competitors nationally and internationally, despite its small size and modest budget. However, it always sees room for improvement.

Innovation in Company F has been concerned lately with the pulp making process. In the production of pulp for newsprint, 80% is by mechanical process, 20% chemical. In the last 10 years Company F has developed a thermo-mechanical process. In its new mill in the UK 60% of pulp will be made by the thermo-mechanical process, 35%

using recycled wastepaper, and 5% chemically. It is possible to use 100% recycled waste, but the paper produced is not of the same quality. This new "recipe" represents an industry innovation. The UK mill will be the second largest mill in Europe outside Scandinavia. For the de-linked pulp the company is using a new technique from France, Germany and Japan.

The second company innovation concerns the yield of pulp from wood. The usual yield is about 50%—that is, two tonnes of wood to give one tonne of pulp. Company F achieves 85% rather than 50% from wood. This innovative achievement was developed in 1983 in Quebec and has led to savings in wood, in pollution equipment, and in energy.

The third company innovation concerns control and involves computerized instrumentation and control. Company F now has a 3,000 foot per minute production rate, and the equipment was developed in cooperation with a Californian company. Company F considers itself to be innovative also in training. In its new mill in New Brunswick it is using simulators like those used in aviation equipment, in training for computerized control.

The strongest motivating factor in Company F's decision to innovate is to be able to compete on a world-wide basis. "If you do not innovate then you fall behind". Other motivating factors are the need to reduce man hours per tonne, and to improve paper quality. These two crucial factors, quality and cost, are now pushing developments in similar directions.

Recent innovations in Company F, although in manufacturing, came from many different roots. Some have been top-down decisions, others bottom-up; there is no set pattern. Senior executives in Company F are open minded, and willing to see and accept new things. Indeed part of their job is to travel the world looking at other industries, as well as paper, in the search for possible innovations.

The speed of progress of innovations in Company F is affected by the organization of the company. Mill managers are responsible for their mills; thus once the mill manager is convinced of the usefulness of an innovation, he becomes the prime motivator responsible for the entire operation. The job of the centre then is to convince one mill manager and, if the idea is good, others will follow. The mill managers are all engineers who have worked through key departments, from mechanical engineering superintendent, through pulp superintendent, and so on. They move round one mill, and then from one mill to another. They are also sent on courses at the best business schools.

Mill managers in Company F have been interested in computerized process control, so acceptance of this innovation has been accelerated. Since computers are reliable, the main point of delay has been on sensors, and difficulties have delayed installations. This problem is common in other process industries. Innovations often require further knowledge and development to be possible. Better knowledge of fibres was thought to be necessary at the moment.

Company F has worked successfully with computers from the 1960s. Sensors have now been developed which, although not the company's own property, have helped to consolidate its reputation an an innovator. Company F was the first in the industry to computerise, and to use twin wires as a way of forming newprint, and has the largest number of new machines of this type. Company F sees itself as managing to stay a couple of years ahead of its competitors.

Company F has a philosophy of promotion from within, which it feels pays dividends in the long-term. This is a long-standing company tradition, which was likened to the situation in Japan. There is mobility around the company, geographically and socially. Salaries in the company tend to be banded and Company F is not amongst the highest payers in the industry.

The degree of latitude in design and specification of innovation in Company F is kept high, but too close specification holds dangers. Keeping dialogue and exchange going is essential to innovation. The company feels it needs wild ideas and does not aim for 100% success throughout the innovation process.

Since innovation is mainly for internal company use, market assessment is internal to the company. Cost assessments are thought important. Costing and evaluation processes are difficult and depend on the project. Company F feel it is impossible to assess the real cost of an innovation. The reward for a good idea may be great and profit generating. Company F's policy is to move in stages to keep costs down, and sometimes it is possible to skip steps. This is not always a successful strategy: for example, in one project by-passing the pilot plant proved very costly because the materials used were not resistant to acid and further work proved necessary. Costing innovation in Company F is flexible rather than rigorous. This is a general rule for R & D projects, otherwise the company spends time and money on costing procedures and not on development. "Flexibility about cost is important because things evolve, and it is impossible to know at the beginning where they will end exactly."

Secrecy in innovation is regarded as important where the company has a two to three year lead time, after that secrecy is not particularly important. It was suggested that because the paper industry is mature and stable there is little exchange of personnel or ideas.

Internal decisions on innovation in Company F involve discussion by the mill, the research group, and the engineering group. If the idea seems good, it is discussed further, costed, and planned at higher levels. Some innovations in Company F come through a reverse cycle, ideas moving from the top down.

The career background and training of the senior personnel involved in innovation decisions in Company F is homogeneous. The manufacturing group is made up of graduate engineers who then attend business administration courses. These people also have much experience in the pulp and paper industry. The president is in practise the only person who could stop any innovation, company accountants could not. Company F operates a post-audit appraisal system after two, three, and five years. This is a formal review which goes to the Board of Directors for consideration.

Individuals acting as champions can be very important to the future of an innovation in Company F, but there are dangers in trying to institutionalise innovation. The respondent explained: "You cannot wait for all the information, you must take calculated risks. You need people at the top who have a positive attitude not conservatives; but creation of a position, say Vice-President Innovation, might even retard innovation".

The general financial position of Company F was regarded as significant for innovation. The exchange rate (particularly that between Canada and the US) was seen to be of great importance to innovation in the paper industry. Adverse variation in this rate can change the company's financial picture overnight. Also of importance in developing innovations were the actions of competitors, and changes in customers' technology, such as printing papers in colour and the increased speed per hour of printing which require better quality newsprint.

For the Company F respondent, the most important factor in being an innovative company is attitude, "Starting from the top, if you do not encourage innovation, you do not get it". A supportive milieu which encourages dialogue between people at all levels is also vital. Third, "People must be left enough time to innovate; if they are overworked day to day, they cannot innovate. They need time and resources".

Company G

Company G is part of a larger industrial group. It is responsible for the group's interests in paper outside the USA. Principal subsidiaries are located in the UK, Belgium,, France, Brazil and India and offices and warehouses are established in some 30 countries. The company also co-ordinates key strategies for group paper companies, including those in the USA. These subsidiaries make a range of high grade printing and writing papers including branded lines. The company is also the largest exporter by value of paper products from the UK.

Although the paper industry has experienced the recession, Company G's group activities in 1982 had shown an increase in turnover of 17% and trading profits were up 60% to £31 million. This was felt to have resulted from operating in higher margin, specialised product sectors, particularly in carbonless copying paper. Although the company has been exporting strongly, it has experienced pressure because the UK domestic market has contracted. Sales volumes were down and although the dollar price for pulp was weaker, selling prices could not be raised enough to cover increased costs in sterling. Improved operating efficiencies, raw material and energy usage, together with fixed cost reductions, more than compensated for this squeeze on margins. The result was a small reduction in the loss from the UK operations.

There were two recent significant innovations in Company G. First was an embossed paper for artificial leather cloth which was developed in a division (R & D provided advice and skills). The product is subject to continuous changes, with an enormous amount of incremental innovation, a procedure which is part of the way of life for the company. The second innovation concerned bespoke papers made to individual specification for customers. These two innovations were used to illustrate the two types of innovation within the company—the first introducing a new product and improving it in response to continuing need for upgrading, the second in the production process, and aimed at significant cost reductions. The Company G discussions centered on the product innovation.

The innovation was developed in reaction to changes in the market-place. The company regards itself as responsive, and has many innovations which start from a problem solving basis, and then are taken further. Company G regards itself as innovative and feels that it has an ideal environment for innovation, with pressure in individual business areas, but the organisation as a whole taking a larger view and allowing opportunities for long-term commitment. This combination of long-term perspective and short-term pressures was suggested as likely to result in interesting innovations.

Decisions to innovate in Company G are affected by perceived market opportunity. In the case of the embossed paper for the artificial leather cloth, an important element was contact with end-users and designers in the industry. The company saw a great opportunity to improve this leather-like look. A business team decided to attempt it and were prepared to continue the search for two to three years. In fact it took about two years to get to the first stage of the project, and then it was obviously going to be a success.

Company G takes a cost benefit approach to any market opportunity, or it follows a strategy which ensures protection of existing business. Embossed papers are an upgrade of an existing product mix, and can be made plainly and cheaply. Involvement with the design end of the business can however make much more money. Company G's embossed product with its greater fidelity to leather is much favoured by designers.

Recognising market opportunity was seen as essential to innovation and vital to the survival of the business. There is no other way that Company G could have continued. The company uses intensive working parties, business teams, to turn the business around. This approach to innovation, and the speed of decisions on innovation in Company G, have been greatly affected by increased competition from companies in the US. Company G now has a business review system. The chief executive takes a longer term look at the business and its strategic plan, possibly twice yearly. The management team meet monthly to take and ratify decisions. There is a team analysis of problems and approximately two months are allowed for preliminary solutions. The business review system acts as a continuous checking process and allows a realigning of plans.

The speed of progress of innovation is affected by the timescale for the evaluation of trial work and the rate of customer feedback. In some product areas this can take a long time. In the case of the leather-look cloth, it was necessary to get greater fidelity of design and Company G had to investigate both the surface to be embossed, and the embosser. Findings designs proved a difficulty and the innovation turned out to be a major logistic exercise. Components of the final innovation package came from Japan, Germany, and the US. The company was forced to go around the world to find people at the peak of their professional competence. The development of the coating technique was subcontracted in the innovation process. The company postpones investment in assets as long as possible, but this builds in logistic and time constraints. If subcontractors used in the initial stage of an innovation are successful, the company is inclined to stay with them.

The R & D contribution to this particular project was primarily on the small pilot plant, and at the trial stage. The technologies essential to the innovation (the extrusion of polypropylene resins) came from a UK contract coater, and from the manufacturer of embossing rollers. This combined chroming technology from the US, embossing technology from Japan (the first spur to the work), and engraving from West Germany. These functions were not available within the industry.

Company G is concerned with high added value, and monitors the technologies of other industries. The company had found little or no development on paper with these coatings. Initially, approaches to the Japanese were unsuccessful since they were not prepared to to supply the embossed rollers. The company then needed an alternative package from the USA and Germany, and this increased the overall project time. Company G feels it is now in a stronger position because of this refusal, but orignally the lack of cooperation was seen only as a constraint. Although the Japanese had been the first in the field with this type of innovation, Company G feels it has now caught up and indeed overtaken the Japanese in Europe, where it is recognised as the leader in design. Company G also made an impact with this product on the American market.

The specification on product quality for this particular innovation was very narrow, but there was a high degree of latitude on how this was to be achieved. Variability is of little interest to the customer, who is only interested in the fidelity of the final product.

R & D and other company functions work closely in harmony with marketing for all innovations. Company G knows the end uses required by all of its customers and it engages in end-use analysis so the opportunities open to anyone who could produce a more faithful design were known. The innovation, partially market pushed, is now a major export business and Company G is exporting 90% of this product. The cost assessment in Company G for this particular project, as for most others, was concerned with the potential return to the business of enriching the product mix. The increase in added value is estimated and the leverage this will give to the business is assessed. This gives an estimate of the rewards likely to result from the success of a project. Although Company G does cost-benefit analyses, the production costs as in this innovation are often uncertain, where uncertainty arose because extrusion was sub-contracted and embossing speed was not yet known. Company G had a good estimate from the debate in the business team of the time necessary for development. A selling price was arrived at by considering the Japanese selling price and making a marketing assessment of what the design was worth. A design is likely to have a life of two seasons, but the value added by the process is very high. At the time of the study plain paper was worth about £700 per tonne and embossed paper of this type about £2,000 per tonne. The large margin between these two figures allows Company G "room to manoeuvre".

Secrecy was regarded as important in developing this type of innovation, although the paper industry generally was said to be fairly open. In this case although there were few competitors, they were all eager to solve the same problem. Keeping the production know-how secret for as long as possible gave a considerable competitive advantage. Design ownership which rests with Company G is vital to the success of this innovation, and is becoming increasingly important in the industry.

The internal decision-making process involved in the innovation centred on the business team, who work via a business review system from the board. The idea of business teams has been adopted throughout the company for all business areas. Company G is a multi-product, multi-site business, and this, it was felt, can lead to problems without a system of this sort. Whatever the size of the business, R&D within the company is managed by the business teams operating in the divisions which are run as profit centres. The business team consists of five or six people, the divisional manager, who is likely to have a physics or chemistry rather than an engineering background, the mill production personnel (the site manager and line production manager), and representatives from marketing, finance, and technical services. The marketing and financial people involved might, or might not, have a technical background. This group then concerns itself with the direction of R&D and with the company's innovative efforts.

Two-thirds of Company G's R&D work is for the main division of carbonless paper, and R&D relies on the businesses to provide ideas and indicate needs. The central R&D group considers it has a role as "provocateurs". Company G's central R&D personnel number about 200, of whom 110 are graduates or above. Development work is done in the divisions. The company has no interest in "knowledge for knowledge's sake", but targets research to lead to changes in processes and products for the company. Target research has a time of two to five years and development work is on a one to two year time scale. Development in about 10 divisions employs about 250 people. Within Company G's corporately funded work there has been a move away from emphasis on developing processes and new process technology and a move towards new product development. The general economic climate and in particular the fortunes of the paper industry have affected attitudes to R&D in Company G. R&D plans are strongly challenged at company budget time, and must be justified on a cost-benefit basis. Even though the economic climate has exerted pressures, the company has retained the integrity of the R&D team.

In Company G a major champion for innovation was regarded as important in the innovation process. In the development of embossed paper the champion was the mill site manager who was most strongly in favour of the project because he saw its importance to his site. The function of the individual champion in the company has been somewhat lessened by the development of the business team; fostering and championing a particular innovation is now regarded as a business team responsibility.

The financial position of the company is less likely to affect innovations than the financial position of the particular business units. The actions of competitors were seen as crucial to innovation in this particular industry, providing an impetus to move forward. Another important factor is good design capability. Although the supply of qualified personnel was generally not a problem in Company G and R&D personnel were regarded as good, the ability to design was viewed as rare.

Management with a significant commitment to innovation and to the development of innovations is important to successful innovation in Company G. Unexpected technical problems can sometimes cause difficulties, but with the right business strategy they are likely to create major catastrophies. Business survival in Company G was seen as dependent on continually improving the business mix. The organisational structure in the company, focussed on the business team, had developed slowly but was now an integral component of Company G's philosophy and corporate culture which was considered to be "right for innovation".

The aim of Company G is to make profit out of technology. They are business and profit oriented on all levels, and over all timescales. The nature of business requires technical change, it is selling into markets where technology is changing rapidly, and it has to respond to this. R&D must act as a change agent within the company, and must help in the drive to reduce costs and develop new products. A close relationship between all of the central functions and the divisions, together with the overall business orientation of Company G, has allowed it to be innovative. Company G executives now have "the correct perspective and keep in mind at all times the nature of the business and business ends".

Company H

The Company H respondent related the pace of innovation in the paper industry to the pace of capital investment. Most innovations take place when a new mill is being built. He noted a tendency in the UK paper industry not to wish to be first in the world, and felt this to be true of all process industries because of scaling factors when bringing equipment up to full size. There has been an improvement in scaling, which has moved from a craft activity to a more scientific one.

For Company H the pulp industry is a chemical industry, and the manufacture of paper is a mechanical process, thus innovation needs to be done by machine makers and by suppliers of chemical additives. The respondent from Company H explained, "The industry needs accuracy and a useful product. It is too wasteful with the high fixed costs in the industry. A major innovation with teething problems can cause serious financial problems". For Company H, success depends on a mixture of people and money: "It is crucial to have inventive and clever people who will find ways of doing interesting things on a shoestring. The paper process is a series of components and one can make improvements all along the line".

The most significant innovation to be introduced in the company over the last 12 years was in on-line measuring systems and the ability to control basic machine parameters. Better measurement and computer control has also given the opportunity to introduce other changes dependent on improved measurement and information display. This project started in the company with equipment leased from the US. The company buys measuring heads, but is building its own. Computerized process control equipment is regarded as a big cost saving when it can be developed in-house and the company feels it is good at this, since it has a very highly talented in-house team available. The other UK company in the study felt such systems should be bought in since the costs of developing them in-house far exceed any benefits. The major product innovation in Company H has been developed in cooperation with a packaging company. This has involved the company in tremendous cooperative developments. Corrugated packaging is particularly important in Holland where most sales are to the open market and not inter-company. Forty per cent of the company's output is corrugated paper board and, including Holland, Company H is the largest producer in Western Europe. For Company H process and product innovations are inter-linked, a machine paced industry can only make better products by running its processes better.

The decision to innovate in the company is affected by profitability. "Capital investment should be for profit improvement, not loss reduction or standing still, so investment must be in profitable areas". The major difference the respondent noted between the UK and the rest of the world in pulp and paper production is the overwhelming importance in Canada and Scandinavia of the paper industry to their economies. Investment, then, is to support basic industry and not to add to industry viability. The respondent felt the US to be more similar to the UK but with better natural resources and cheaper money which, in capital intensive industries, can be a most important factor. "In basic industries 7% on sales could be regarded as a good profit and a 10% to 20% return on investment is a good performance. This only remains good with low interest rates."

Innovation for Company H was dependent on the availability of ideas and funds. "Everyone is always looking for new things, with innovative people at all levels you do better, but such people are hard to find". Big innovations in Company H tend to be linked to big expenditures of capital, " . . . inevitably in industry old plant gives way to new". Although investment was seen as necessary to be cost effective, it was thought possible to exist on old equipment for a long time. It was suggested that the cost effective life of plant if it is subject to regular and extensive updating is about 70 to 80 years.

In Company H the speed of innovation depends on the scale of change and its implications. Development in paper making is seen to a large extent as an adaptive process which starts with the existing production line. It is possible for Company H to change easily—if there is a small good idea, it can be done at once. The next level of change is concerned with parts of the process and can take six weeks to six months. Significant here are machinery lead times. If any change to a piece of machinery is anticipated, the time taken by the change is related to cost. Items costing £100,000 can be changed in nine months, and spending up to £130 millions can happen within two years. The time involved in change is also affected by constraints common to all process industries; major changes tend to take place in holiday shut-downs because of the high fixed costs.

Necessary technologies for the recent innovations in Company H have come from the company's technical and R&D departments, and from suppliers. Paper and board, it was suggested, is an open industry because of this supplier orientation. Measurement and control systems have been developed by the electronics industry. The company can now measure almost anything on-line. Process control in-house is thought to be very good.

Company H has at times experienced problems with the adequacy and speed of supply of key items and personnel, but has not been dramatically affected. "These things are never really a problem during the recession. A queue of people are waiting to supply everything that one might need. Personnel are never easy to find, that is really good ones, but we have managed to do this."

The degree of latitude in the design and specification of innovation in Company H depends on the kind of project. In additive projects, if they do not work you switch them off, and the degree of latitude therefore depends on costs and likely profitability. In absolutely fundamental projects involving major investments, the company is not prepared to be innovative; it wants to be first. The risks associated with any extended shutdown are too great and Company H regards such risks as too costly. For Company H, paper is an incrementally innovative industry, very dependent upon machinery suppliers, and in the machinery industry there has been an enormous amount of change. The predictability of machines is now much greater, but it probably still takes years to reach peak performance on site.

The company engages in market forecasting. A continuing major brief is to look at consumer industries, to anticipate and meet their needs by adapting products to perform better, or by producing products at lower costs. New uses are not the company's major concern, since they are rarely selling to the consumer but are selling mainly to printers or convertors.

Cost assessment for innovations in Company H tend to be no different from those involved in the ordinary budgeting process. The aim is to evaluate cost structures against competitors, and to look at major points of uncertainty such as exchange rates. "These are still a nightmare and have been for some time". Enormous exchange rate swings can make or break competitiveness. Company H feel that the major requirement is for stability.

Within Company H secrecy on product developments in-house is important, but major processing developments, and cost related developments do not depend on secrecy. The industry was thought to be very open, ideas are circulated, and this was regarded as essential.

Decision making on innovation in Company H relates to the cost and size of the innovation. The industry was viewed as R&D based but Company H's production and buying personnel can also decide, and implement change. Capital investment decisions are tested on an executive team of three, all of whom are on the board and have engineering backgrounds. The financial director, who was originally an accountant, contributes to this discussion. Those involved in major decisions in Company H are mostly engineers, although decisions about innovation also involve accountants and sales personnel. All the people involved have been in the paper industry for a considerable time and many have grown up in the industry.

A major champion for an innovation in the company is important. The company is a decentralized organisation and the mills are run as separate businesses. "The best innovations are pushed through by those with a burning bee in their bonnet. Fanaticism pays in this area".

Both the general economic climate and the financial position of the company were of significance in contributing to its attitude to innovation. Problems of finance for particular schemes were of minor importance and secondary to the wider economic and financial outlook. The actions of competitors were seen as important, as were the requirements of customers.

For the Company H respondent, a company in the paper industry requires talented people with proven excellence in engineering, chemistry, physics and production skills, and money, to be innovative and successful. The government's role should be to ensure that competition is fair, rather than to intervene frequently. Incremental rather than radical innovation is required.

THE MINING AND METALS INDUSTRIES

Generally weak industrial activity in the world's major economies and the delay in the awaited upturn which has been expected have severly affected activity in the mining and metals sector. High interest rates and the low price for most metals have further exacerbated theses trends. Future improvements require movements in metal prices, exchange rates and interest rates. An improvement in international business confidence could restore the activity level in the sector.

Company V

Company V's basic strength is in primary metals and is founded on the company's ore reserves. The company markets many different forms of nickel world-wide. Its largest consumer is its alloy company, an international group producing rolled, forged and machined products for the energy, chemical, aerospace and other major industries. Company V also has a variety of other interests. At the end of 1983, the company employed about 25,000 people in 21 countries and had total assets excluding ore reserves of some $3 billions (Canadian).

The past years are regarded as having been some of the worst on record for the mining and metals industries. Demand and prices for non-ferrous metals including nickel and copper declined sharply as the world-wide recession continued unabated. At the same time, unit production costs continued to rise, reflecting higher costs for labour, energy, and other expenses, as well as reduced levels of production. Both of the company's major operating segments, primary metals and alloy products, were severely affected by the depressed markets for their products. The company has suffered losses for the past two-and-a-half years, and the market still remains depressed.

The company's recovery will be dictated by the timing and strength of economic improvements in the industrialised nations. The success of the company's efforts to reduce costs could be another important factor. Company V's basic long-term objective remains the achievement of profitable, sustainable growth in businesses involving technology, broadly compatible with company skills.

The company is structured to reflect main business interests. It is concerned with primary metals and alloys and, secondarily, with some small business ventures. At one time Company V's corporate R&D had three laboratories. They employed 300 professional and technical personnel and there were more R&D people in the business units.

In the 1970's the company established a diversification programme of which one component was to develop new process operations through R&D. Company V was prepared to take risks in research because it felt it could afford to wander from the main business thrust. The new business ventures area started several small businesses, but the onset of the recession and other difficulties re-focused Company V on its primary lines in the metal business and ultimately led to the closure of the corporate R&D function. There remains an R&D establishment for the primary metals business, which is concerned mainly with process development and an R&D function for the alloy products which is less multi-disciplined than it was and has been proportionally reduced. Although the company still does research, it is now more directly related to existing lines of business. The transition has been difficult.

Innovation is still considered vital in Company V since it is seen as the principal way to get new products and processes that are necessary for competitive advantage. As most facets of the metals business are relatively mature and frequently threatened by substitutes, proprietary niches are actively sought to ensure competitive advantage. In the metals business, it was suggested, technical advances are coming faster, and technology has a greater role than it had in past decades, therefore for a company to be successful it now needs aggressive technical activity.

Justification of research programs is usually based on an assessment of the available market in alloy development or on process needs or potential productivity improvements in the primary metals area.

The decision to innovate within Company V comes from the people doing research. "We provide the context in which they can work and create the right climate to innovate, that is the trick of management." Company V is inclined to discuss results not ideas. Innovators must prove their point. If there are two or three possible projects, an analysis is made of the likely project return and the project is selected in this way. The importance of not choosing too soon, but soon enough to leave resources for development was stressed. The universal natural inclination it was felt, was to avoid big risks but to be willing to take many small ones, this was not regarded as a good strategy. Decisions about innovation in Company V can take anything from months to years. Those who do not like risks shelve more ideas. Risks can appear to be minimised by testing ideas, but the company may get more tested than it wants. The aim of Company V is to encourage even the conservative to go out and try.

The main factor affecting the speed of progress in Company V, once an innovation has been adopted, may be merely money but may also be a function of related technological development. A recent innovation in Company V was made possible by fundamental advances in materials science, which allowed engine design advances. Equally important were advances in analytical equipment.

The Company V respondent suggested that although it is possible to be specific about design on innovation, sometimes, the end result is completely different. If people are working with freedom, unexpected but useful products may arise. As long as these are in touch with the business and what goes on in it, they must be regarded as a very useful result. Too specific targeting can be a severe constraint, and Company V recognise the need to be flexible.

Company V is constantly assessing markets and the actions of competitors. A thorough cost analysis and likely profit estimate follows when a product is considered commercially realistic. There is a tendency, however, as in many companies to confine itself too much to traditional markets and to take too short-term a view. It is probably necessary to look further into the future. "Today's market is yesterday's market, what we need is tomorrow's."

Secrecy is relatively important but probably not crucial to developing innovation in Company V and in the industry. Wherever possible, patent protection is sought for significant developments in both the process and product areas. Know how is also recognised as having a value when it allows for a time advantage on the competition.

Once an innovation has been taken to the pilot scale, in Company V then a business unit will take over. Technical people, it was thought, can only take a scheme so far and then it must move into the business context.

The people concerned with innovation decisions in Company V are initially technical, then business people generally with a technical background, but occasionally a financial background. A major champion for an innovation was regarded as very important. The ideal person, it was suggested, is one with sufficient influence to enable him to carry out whatever was wanted. If the champion is in R&D, the view of R&D in the company becomes vital. Without a champion, it was suggested that the odds are against innovatory success.

In Company V shorter time horizons in R&D have led to innovations which are smaller, and the respondent felt that these time horizons must be enlarged somewhat. Support for innovation, it was suggested, needs to come from the top. The main problem was seen as balancing long-term needs and business relevance which can be extremely difficult.

The general financial position of Company V was considered of very great importance to the likelihood of successful innovation, as were the actions of competitors and market changes. A supply of qualified personnel was regarded as a necessary, but not sufficient, condition for innovative success, and the supply of key materials or components could often be crucial. The correct structure was also considered of significance.

Environmental regulations in the mining and metals areas are of significance as they are in Company R, since they put a major call on resources. Unexpected technical problems, although only significant occasionally, can completely sink a scheme, but for the most part Company V can overcome them if there is willingness not to give up on the project. Unexpected cost changes are also very important when the costing on a project is close. The influence of customers can be decisive.

Company V feels that for the industry, it is as innovative as anyone else. It has a commitment to R&D, to innovation, and to developing its own new opportunities. Success in innovation was seen as requiring this consistent commitment and adequate resourcing. "What you need is to create the right environment for innovation to flourish, but this is very difficult to do."

Company R

Company R and its subsidiary companies form a British-based international group of mining and industrial companies with interests in almost every major metal and fuel, including aluminium and its products, borax, coal, copper, gold, industrial chemicals, iron ore, lead, oil, silver, special steels, tin, uranium and zinc. The company's interests throughout the world are in organisations varying widely in size and degree of development.

Company R's headquarters in London provide services such as consultancy, environmental advice, insurance, legal, planning, purchasing, secretarial, taxation and technical advice. The group's profits have been affected by the generally weak industrial activity in the world's major economies, and the delay in the improvement which had been expected to take place. High interest rates, particularly in the US, were a major contribution to the lack of recovery, and to the low level of prices for most metals. The company's prospects are seen to depend to a considerable extent on the movement of metal prices and exchange rates. Drops in interest rates and weakening of crude oil prices should help reduce inflationary pressures which suggest in Company R that an upturn is more likely to materialise. Metal prices would respond quickly to an improvement in international business confidence and Company R could benefit accordingly.

Company R does not regard itself as a high technology company; it runs 30 mines and smelting complexes throughout the world and its main search is for cost reductions. Most new developments are aimed at this. The latest development in Company R which saves millions of dollars is the replacement of a 200-tonne diesel driven truck by an old-fashioned trolley. A second development is concerned with new ways of making steel. Company R sells iron ore to Japan and Europe and new ways of making steel are clearly of great significance to it. The aim in all innovatory activity in Company R is to be better than the next company, but it has no interest in innovating outside a known business area. Decisions on innovation in the company are quick, taking a matter of months. The final decisions are made by a committee of the board which is chaired by the chief executive.

In Company R the progress of an innovation is not seen as a market problem; it is developed as fast as the company can put it together. The company rarely experiences problems with the adequacy or speed of supply of personnel, since it pays very highly. It regards this as an American approach, an attempt to buy the best, and goes only for top class people.

The latitude in the design or specification of an innovation in Company R could be very wide, and a decision would be taken by the design team and project manager. The company is not concerned to make market assessments but concerned with innovation to improve methds of work. Cost assessments, however, are always made and are concerned with standard internal rates of return. Secrecy is not considered to be important in developing a new innovation in Company R.

The internal decision-making process involved in the development of an innovation can be from the bottom up or from the top down. The trolley idea arose at an internal symposium and subsequently went forward to the board for ratification. It was suggested by the chief engineer. Personnel involved in innovation decisions in Company R are, for the most part, professional engineers with 20 years experience in the industry. A major champion for an innovation was regarded as of some importance, since it was felt necessary to have someone with push and drive to move an idea forward.

For Company R the ability to innovate is affected both by the general economic climate and by its own financial position, although acquiring funds for a specific project is generally not a problem. The actions of Company R's competitors are not usually seen as likely to cause worry, but unexpected technical problems may delay an innovation.

R&D in Company R is regarded as an activity to improve the technologies used in existing operations, and to provide improved technologies both for existing operations and as a basis for new operations and diversification. The corporate structure of the company, and specifically the fact that major subsidiaries are often less than wholly owned and have considerable autonomy, means that each group company is responsible for undertaking the R&D it needs to achieve its business objectives. At the corporate level, the R&D function is limited to maintaining general awareness of technological development in the subsidiaries, sponsorship of university research relevant to the corporations interests, and development as a basis for the diversification of inventions and innovation brought to the corporation. Company R does in-house R&D wherever it forsees commercial advantage being derived from successful outcome. The company no longer attempts to quantify returns from R&D investment, but programmes are evaluated at frequent intervals and abandoned if commercial objectives are judged to be no longer attainable. Company R's major R&D effort is currently concerned with improving technology in established operations and processes to extend established product ranges.

In the operating companies the main thrust of R&D is determined by the chief executives, generally on the basis of discussions between marketing, production and technical directors, and managers. At the corporate level, decisions are made by the chief executive based on recommendations from his head of research and other technical advisers.

The main objective of Company R's innovatory policy is to be more competitive by becoming more efficient. The company regards itself as being as good as everyone else, rather than highly innovative. It employs high-calibre, multi-disciplinary groups, and sees itself as cost-conscious and efficient rather than innovative. Recently, since Company R produces lead, it has decided to develop its interests in batteries and has bought a company to ensure it is up to date on the way that battery technology is changing.

It has also bought a Russian lead smelting process which is thought to be the best in the world. Some of Company R's subsidiary companies are regarded as very innovative. One in particular, concerned with double glazing, is innovative in new product development; its latest product, which is a market research led idea, is to meet a demand rather than a need. Company R also engages in cooperative ventures to increase demand for copper, aluminium and zinc alloys and so on, which it regards as innovative.

The Company R respondent felt that most of the processes relevant to the industry are well known and understood; cost cutting is the only innovative option open in mining and smelting and there is constant pressure on the business to be more efficient. There have been many new developments in the company. Smelting, and zinc innovations have come from the operating side, and have then been assisted by R&D, whose funtion is to help in this process, and to keep the company the best in the world. The Company R respondent did not feel there were any great differences in innovation in the metals field in the US and the UK, but suggested that US companies search more consciously for new products and have a more organised approach to innovation. He regarded US companies' market orientation as stronger and praised the quality of their personnel and the orientation of their senior management, and suggested that they could have a major impact throughout the business.

THE PETROLEUM INDUSTRY

Overview

The oil sector contains many of the largest companies in the UK and the USA. Fifteen of the top 50 companies in the USA and seven of the top 50 in the UK are in this sector. In the UK, all of these companies except one have major interests in other countries, including the USA, whereas in the USA there is a larger number of companies with a more national orientation. Many US companies grew up around the local oil fields of California and have expanded from that base, whereas there were no such indigenous resources in the UK until the discovery of North Sea oil in the early 1970s.

The companies in this sector are major spenders in complex businesses in changing market conditions. In the mid 1970s, with the threat of oil supplies drying up, finding "substitutes" for oil and natural gas became a matter of urgency. There was great interest and investment in such technologies as coal gasification, catalytic crackers and the application of biotechnology to produce new energy sources. In the later 1970s and 1980s changes in the price of oil and in the energy scenario, together with the downturn in the demand for chemicals, led to attempts to diversify the business with an emphasis on new ventures and high technology with a strong research bias[52]. In addition, the urgency to find substitute oil and gas products abated and it is now widely perceived that there is less scope for a global technology in the 1980s. Each country is trying to exploit its own energy resources and to accommodate its own political prejudices in making the best of an era of expensive energy.

Demand for oil peaked around 1980 and since then demand has been greater than production which has been constrained deliberately to reduce inventory. Both the gas and oil businesses are highly competitive and easy availability of crude oil has led to low petrol prices. There is every incentive to improve productivity. The oil companies, although they are still able to maintain high profits, are facing rather more difficult circumstances than formerly.

The energy industry seems to be undergoing fundamental changes. The relationship between energy consumption and economic activity seemed at one time to be one of parallel movement. Increases in gross national product are no longer accompanied by comparable increases in energy consumption. Although it seems likely that energy needs for the world will increase, the next few years might be somewhat uncertain.

Company O

Company O is based in the UK but has world-wide interests in gas and oil. It spends approximately £300 millions on R&D. This amount has increased from £222 millions in 1980 and was considered by the group coordinator for R&D to represent an enormous increase.

---

52 Swords-Isherwood, 1983, Op.cit.

Company O was one of the first companies to establish a customer-contractor principle for running its R&D in the late 1960s. In 1982, however, its R&D effort underwent a major re-organisation, cutting about 800 research jobs, one-fifth of those employed in its central laboratories. This was partly the result of automation of much of the drudgery of research. The company aim was to have a new research programme involving fewer staff but to maintain effort and allow more freedom for speculative research.

Company O tries to be innovative both in terms of major radical innovations and in terms of incremental innovations. Its last major innovation was concerned with the undersea manifold for the production of oil and gas. It has also been working on coal gasification.

For Company O, cost effectiveness is fundamental to its innovative approach. It also sees a need to bring new products and processes to the market. The procedure involved in innovation in the company depends on the type of innovation. If the innovation involves a capital programme, Company O calculates in terms of the earning power of a new process. For new products the company determines the the size of the market and whether the investment can achieve a positive net discounted value which would make the innovation worth undertaking. All projects are closely costed. The company feels this is possible despite the speculative nature of R&D efforts and the risk involved. Its method of costing R&D has been developed over time. The timing of innovations in Company O is very variable since innovations concern a wide range of activities. Timing can range from six months to five years.

Fundamental to the innovation process in Company O is an analysis of the market-place and an ability to determine the place of a new product or process in it. Costing is also considered of great significance in this process. Nearly all the technologies used in recent innovations in Company O have been developed in-house, though the company will buy in any necessary technologies and has done this in the past.

In Company O the process of innovation and decisions about innovation are effectively a team operation. There is an inter-play between marketeers, manufacturers and research people. The whole is a business oriented exercise. The decision to go for a new innovation has always to be dependent on agreement being reached amongst all the parties. The danger of this strategy, it was suggested, was that a lowest common denominater result might be achieved. But Company O makes strenuous efforts to avoid this. Depending on the nature of the innovation, the company might use its own marketing skills, or if it is a very new development, it might use a good market research organisation. The company does not favour one line approaches, but attempts to make a range of approaches to the market.

The nature of the costing of an innovation depends on the type of innovation and on the point in time during the process of innovation at which the cost estimate is made. If Company O conceives of a new idea which is market oriented, the costing may be very difficult because the company will not know whether the idea is technically possible. If the innovation arises in manufacturing or production and is pushed by researchers who think they can do it, the company will expect the costing to be fairly tight; but in essence, the Company O respondent felt that costings of innovation were very much judgemental on a large scale. Development, it was felt, was not so hard to estimate. For new ideas in research, the first technical aspects in the innovation process, although they are of fundamental importance in the business, it is difficult to predict their outcome and costings are of necessity uncertain.

Secrecy in innovation was regarded as of very great importance in Company O. If the company acquires a technical lead, this may enable it to steal a march on its competitors and to patent an idea. If it is prevented from doing this, then initial efforts on the innovation may be wasted.

Innovation decisions in Company O are business decisions, they are not made by manufacturing, by marketing or by research, they are made by the business which these functions serve. The company is organised on a business centre concept, rather than functionally. It was restructured a few years ago from a functional organisation. The Company O respondent liked the new organisation better since he felt it overcame more problems than the old system and created very few. The business centre manager will either agree an innovation or, if necessary, he will put it forward to the next level of management. In the past, when Company O was organised functionally, insufficient emphasis was given to the good of the business. It was suggested that the dangers in the present matrix organisation lie in the possibilities of arriving at the lowest common denominator and of insufficient time being given to work particular projects through.

On balance the respondent still felt that a matrix organisation was a better solution.

The background of the senior management involved in the innovation decision is variable. In technical areas in Company O there are marketeers who now research or manufacture. The company is well mixed and aims to continue this type of mixing. Thus the manager in charge of an innovation might have any sort of background—he might be an accountant or a lawyer. Because of the nature of the business and of the group, he is most likely to be an engineer; however, Company O does not assume that engineers make better decisions. The system for reaching a decision on innovation means that any decision is likely to have had a multi-disciplined input, whatever the background of the person who finally decides.

A champion for an innovation was considered as very important in Company O, although the respondent felt that there was no ideal champion. The person who has the idea must push it on. It was suggested that in any organisation it is necessary to have someone who wants a particular innovation to survive and who is prepared to fight hard for it.

For the Company O respondent, an innovative company is one in which the management has its eyes open and is looking to solve problems of tomorrow, five years and 10 years away. The management must be prepared to support the cost of determining how to operate in an uncertain future. Top management must " . . . put itself on the line on this, and must set the tone for the rest of the management team". Investment must be on an on-going basis, avoiding any switch effects which lead inevitably to loss of confidence.

Company P

Company P, although only medium-size for an oil company, has a turnover of $10,000 millions. The company, which is US based, is organised into seven product orientated divisions with one division for exploration and development, and one for R&D.

Company P has not chosen to become a conglomerate. It has, however, branched out into mining and has bought new operations in the chemicals area. Nevertheless its major effort is in petroleum and gas and its latest innovations have been in this field. The company has just seen a major innovation through to its exploitation phase with the production of the first shale oil. The success, or otherwise, of this innovation will be demonstrated in the next few months.

Employment within Company P has remained steady. The company has a definite policy of no lay offs but has not been recruiting staff. It has been hit late by the recession during the first half of 1983.

Company P spends around $60 millions on R&D and the respondents considered that this, although average for the industry, was large for the size of the company. They suggested that the oil industry as a whole spends a low percentage of turnover on R&D. Company P's R&D centre is organised into departments by product that more or less parallel the divisions of the company. Fifteen to 20% of the centre's work is done on a customer-contractor basis, the bulk coming from corporate funds. The respondents felt that there have been many innovations in the company from all divisions. Most innovations start in-house and generally involve either new chemistry from which the company develops new products, or new processing techniques. For the company to proceed with an innovation, it must be relevant to the company's brief. Innovation requires creativity but also requires awareness of customer needs. The R&D centre has its own technical sales department which is responsible for selling ideas which the company feels are outside its central remit. Projects are allocated priority according to their value to the company, and the speed of development of a project depends on its priority. Company P allows flexibility within a project provided that it fits into the framework of company operations.

Company P believes that a certain amount of market assessment is necessary in innovating before embarking on a development phase. After a basic idea is conceived the handling of an innovation is decided by the market. Each project is evaluated by broadly the same criteria and Company P uses returns on investment to assess projects. It was agreed that it is very difficult to cost new projects accurately.

The respondents from Company P believed that secrecy is important in innovation. Patents can cover major concepts but the process is protected by maintaining secrecy on know-how.

The innovation process within Company P follows a standard pattern. The idea is tested by describing it on a concept sheet which is submitted to the patent department and to a committee of senior managers. It is then ranked as worth proceeding with, or as needing more laboratory work before a decision can be made, or as not worth pursuing. Then innovation is an iterative process with check-points all the way through. Monthly reports are generated with the operating plants which, in turn, constantly carry out market research. The final decision comes when a project is at the pilot plant stage because this requires large capital outlay.

Environmental regulations have affected innovation in Company P and in a few cases may have prevented projects from proceeding. Generally, their effect is to add cost to the project.

The respondents from Company P felt that the recession has affected the company across the board and that the general financial position of the company is relevant to innovation.

There has been greater competition within the company for new projects, but there has been commitment to innovation at the highest level and the R&D centre is fully aware of the support it receives from the president of the company. Despite the recession, commitment to R&D has resulted in a major new innovation coming on stream. Personalities in Company P have had a major effect in deciding the direction of the company's innovation.

Company Q

Company Q has divisions and affiliated companies operating in the US and 80 other countries. Its principal business is energy and it is involved in exploration for and production of crude oil and natural gas, manufacturing of petroleum products, and the transportation and sale of crude oil, natural gas and petroleum products. It also has interests in mining, in sale of coal and fabrication of nuclear fuel. A group chemical company is a major manufacturer and marketer of petrochemicals, and Company Q is also engaged in exploration for mining and minerals other than coal. An affiliated company manufactures, markets and services a broad line of industrial equipment. The company conducts extensive research programmes in support of these businesses and provides capital to innovative new ventures, some of them not related to those existing businesses.

The continuing world recession has been a major factor in the 13% decline in the company's earnings and the company suggests that these results also reflect changing patterns in energy consumption which have depressed oil demand in recent years. The most notable of these trends has been the move towards more efficient energy use and the substitution of alternative fuels for petroleum. Company Q sees these new patterns as likely to persist regardless of swings in the economic cycle. The depressed demand for petroleum products has aggravated a large world-wide industry surplus of producing and refining capacity and distribution facilities, and this had led to weak crude oil and natural gas prices and has placed severe pressure on profit margins as a result of intense competition.

The company is very concerned with innovation and its stimulation and in matching its R&D goals to company goals. The respondents suggested that the role of the chief executive, his beliefs and desires, and the corporate culture that he establishes are absolutely crucial in maintaining innovativeness. Thus the major job of an R&D manager is to convince the chief executive of the technical needs for innovation. It is not necessary for the chief executive to be an innovator, but he must permit it in others. Company Q has isolated innovative efforts; it is concerned with high technology and has been involved with growing small businesses for over 20 years. Crucial to the success of these ventures is the timeliness of management actions. A thorough knowledge of market demand and an ability to keep in touch, are vital if the company is to make money.

Company Q is made up of very diverse businesses, and 40% of total R&D is slanted away from the energy and oil business. At the upstream end of the petrol business, the company has recently installed in the Gulf of Mexico a major structure which is taller than the Empire State Building. It is to be used as a drilling structure in a new field and allows drilling in areas in which it was not practical before. In the North Sea the company is operating below water submerged production systems. Both these projects were huge gambles, because it is only possible to pilot the project to a limited extent. The company spends about $9 millions (US) a year in this area and is also engaged in drilling in the Arctic. In these activities Company Q would regard itself as similar to other oil companies.

Since petrol is very valuable, it is possible to spend a large sum on such gambles: "The prize is worth it". The respondents compared their attitudes to innovation with that found in pharmaceutical companies; for both, innovation is a matter of do or die. Choice, then, is limited since oil is found where it is difficult to work and it must be got. The company has interests in the synthetic fuel area and in the liquification of coal, but it is not pursuing them on a commercial scale. Company Q is concerned with looking ahead 20 to 30 years. At the moment it is spending $100 millions a year on this which might taper to $70 million if circumstances change. If the costs of research in alternatives becomes comparable with the costs of extraction in difficult areas, then decisions on R&D spending would obviously change.

Decisions about an innovation in the oil industry can take a very long time. The projects are very big with very high costs. Fifteen years, for example, is not an uncommon time for the development of a major project. Company Q is concerned at the moment with innovation in process technology, such as flexi-coking which converts crude oil and other waste to useable gas and liquid. This innovation has been licensed to a Japanese oil company which was given a variety of technical help in developing the project. Company Q became interested when the project became more successful.

The speed of progress of an innovation in Company Q can be slowed by the difficulty of taking high risk decisions. There is a need for the innovation to be tried within the company and a need for internal marketing to ensure it is spurred within the company. A continuing constraint is the shifting view of the nature of the market.

In its upstream technologies the company is dependent on work in the materials area and in geology and geophysics. The knowledge base is in Company Q's production research company which employs earth science personnel throughout the world.

The company also needs large numbers of mathematicians and experts in engineering technologies. For the refining and chemicals end of the business the concerns are with process production and sciences, chemistry, mathematics and engineering. Physics, chemistry and engineering are required across the board in all Company Q's business areas. In the company's new office systems development, expertise in electronics is necessary.

The impetus for innovation in Company Q does not lie in patents but in providing personal growth and money for the entrepreneur. The difficulty of keeping a company, especially a large one, innovative in all areas was recognised. Company Q is decentralised and in the areas it regards as important (that is, in the upstream and refining end of the business), it sees itself as very good. The company has a major research effort in a variety of areas, and innovations might come from any of these areas of interest. In 1968, the company established a corporate research laboratory, which although expensive, has produced interesting and innovative ideas. It was originally designed to study science in areas of interest to the company and is funded corporately. It differs from the other company laboratories which are area specific and commercially oriented. One success of the corporate laboratory concerned refining technology, the chemistry of sulpho-dioxide and carbon dioxide contaminates, and how to remove them with new chemical reagents. The laboratory developed a new theory, and new chemicals. These were then transferred to a company divisional laboratory and refinery. The idea is now in use all around the world. The corporate laboratory is concerned with pioneer research, is kept in touch with the business needs, and spends about a $100 millions per year. The Company Q chemical company has had 22 new innovations in the relatively recent past, of which seven have come from the central laboratory.

For Company Q, recruiting the right quality of personnel can be a limiting factor. Key technical individuals are hard to find. On the whole money is always available, and it is very unlikely that an innovation would be held up by a shortage of funding. On occasion, to answer science questions, the company is dependent on state-of-the-art equipment, and this may prove a constraint. In Company Q concern with market assessment is less than in some other companies, but this is partly because most innovation is in process, not product areas. The company is not quite as market driven as many others, but it is concerned with the market for its products. Company Q sees itself as not good at product innovations, although it is attempting to move in this direction, particularly in the chemicals area. The Company Q respondents saw this as a weakness particularly in the future for the new development area which would be more dependent on new products and on market orientation. Company Q is now employing key people in these areas, recruited from companies who have always had a very strong market driven approach, for instance IBM.

Company Q has been concerned with and is very good at assessing costs and future returns in the oil business. Estimates always have as a first step a consensus view on the bounds of oil prices, which are then forecast into the future. This provides the base for calculations on technical innovation. Any technology forecast is then costed, using this framework, the estimated projection of crude oil prices is used as a target for cost and time calculations. Starting projects is always difficult, and sometimes the company pursues things too far, but on the whole the cut off is where it should be and the company benefits from the work which has been done.

Secrecy can be important in innovation in Company Q in certain areas, such as sysmic technology in locating oil. In other areas such as coal liquification, secrecy is not so important as the work is of concern to government, other companies and public utilities.

The internal decision-making process involved in the development of an innovation for the most part takes place in the divisions where an R&D or technology plan is developed. As a project progresses, it will be reviewed by a committee of R&D and business people, and top management will be involved in the scheme at some point. Involvement from the top is dependent on the scale of the project. When it is a big project, the management committee and chief executive are involved whereas a smaller project can be decided further down.

The career background and training of senior personnel involved in innovation decisions in Company Q is mostly in engineering. Amongst the top people in the company at the moment, there are two financial personnel and a lawyer; the rest are engineers. Five out of seven have long careers in the oil industry. Company Q recognises that even in its top executive group it is not strong in the area of marketing expertise, nor are there any top executives with direct R&D experience, although R&D is recognised as critical to the business. R&D is part if the business and is there to serve the business.

A major champion for an innovation was considered absolutely crucial. The respondents stated that nothing happens without such a champion. Management problems and the need to reorganise and motivate management was considered a critical issue. "The top guy is an important controlling factor on innovation. You need a good chief executive and luck." Another internal factor significant in being an innovative company was the ability to ask the right questions, such as "Is your innovation really needed?" All the easy things have been done and so there is a great economic premium in being successful in innovation now; thus within Company Q the driving force is always there to go out and find more oil and process it more effectively.

III. A Traditional Manufacturing Industry

THE MACHINE TOOL INDUSTRY

The machine tool industry has always been susceptible to economic fluctuations and in each country tends to face periodic heavy demand when there is re-tooling by automobile and aircraft manufacturers. When these cycles end the industry can face a severe slump in demand. Cigarette machine manufacturers are also subject to cycles. Technical change in the industry has been rapid and its nature has changed. Product engineering strength remains an important aspect of competitive success. Generally Japanese machine tool firms have been most successful in commercial re-orientation to new markets as traditional ones have declined, and in increasing new markets as new technical opportunities became available. Automation has been the most important technical development in the machine tool industry in recent years. Automation offers competitive advantage through increased flexibility, reduced stocks, and higher machining reliability, but is expensive, and payback time may be long and require long-run objectives and planning from the firm.

Company S

Company S is a UK based cigarette and packaging machinery manufacturer which is part of a large multinational group. It has an annual turnover in the region of £120 millions, improving results from the cigarette machinery business being offset by increased losses in the packaging business.

Company S has a design and development department, rather than R&D department, which employs graduates, technicians, designers and draughtsmen. There are two routes by which one can become a designer; one is for university graduates and the other is by promotion from craftsmen. The company also employs mathematicians and engineers to work on noise levels. It provides the best equipment, most of it microprocessor based for its D&D department. The company is now researching into laser technology.

Company S has three of its own patent agents. These agents have taken out around 1,000 patents, and have also provided information on competitors actions in this area.

The company is engineering orientated but technical problems and needs are defined by the market. It has a Product Planning Group to look for opportunities for the company. This group made up of five engineers and a salesman with full support, looks at the market, at customer needs and at competitors' actions. It interacts with the sales group, the group managing director, and customer contacts to decide what is needed. If a new machine is required this consultative process results in a highly specified outline which will be the basis from which the next generation of machines will be developed. The Product Planning Group is looking a minimum of three years ahead.

In designing cigarette machines the aim is to make machines that produce increasing numbers of cigarettes per minute (CPM). From the turn of the century to 1960 there was a steady progression in speeds attained by machines. Since 1960 there has been an explosion in speeds attainable by Company S's machines. The two major European firms with which the company competes have made comparable advances in their machines.

The first breakthrough was to 4,000 CPM in 1964. This was achieved by changing the way in which the cigarette was made and the tobacco put together. This saved a few percentage points of tobacco which represents a great deal of money and allowed for return on investment on the machine within 12 months. The next development was a machine for single stick cigarettes. The latter became the highest volume product in the world. Subsequent breakthroughs resulted from the use of electronics and new materials. Advanced electronic devices have led to a degree of automation.

Company S's latest development is considered by the company to be jumping a machine generation. The old machine was producing 5,000 CPM and competitors are working on 6,000 CPM. The latest Company S machine produces 8,000 CPM and also produces a better quality cigarette using less tobacco. It is also very quiet, with less than half the noise level of previous machines. Its development called for innovative engineering and involved all the research engineers and all the top executives at Company S.

The development of this latest machine resulted from the use of microprocessors which broke down major technological barriers which existed even three years ago. Microprocessors were used in design and analytical work and the machine incorporates 22 major innovations. It also uses new and more sophisticated materials. It is now under test in the USA but has encountered teething problems.

The machine, covered by many patents, is of an inventive design and is also produced by a greatly improved process.

The Company S respondent felt that all innovations were really incremental, that any leap forward had to be substantiated by much incremental work, and similarly, incremental work could be of sufficient significance to appear as radical innovation. No company, it was suggested, actually looks for innovation *per se*, rather "it is the improvement in the product or the process" with which it is concerned.

The company suggested that its innovative effort relies on good, free thinking, capable engineers of a very high calibre. These people must have a flair for innovation and therefore are difficult to find. It is also necessary to put these people in "an environment of innovation" where they are able to think imaginatively within defined channels. Company S involves its top innovative thinkers in "brainstorming" sessions when it has problems to solve.

Company T

Company T is one of the largest US based machine tool companies. It was chosen for study because Company S, although a cigarette machine manufacturer itself, suggested that a machine tool company would be the most suitable available US pair. Company T's turnover was around $760 millions in 1982, which was a decline of 19% on the previous year. The machine tools industry, including Company T, has been particularly hard hit by the recession as orders from most segments of manufacturing, its main customers, have slumped markedly. Many of Company T's US competitors have quit the industry, merged or radically restructured. Company T, sees itself as different from its rivals in three ways. Firstly, it is broader, being concerned in other materials as well as metals, and producing its own control systems. Secondly, it takes a more positive attitude than its US competitors in the face of Japanese competition, and rejects the public image of the machine tool industry as reactionary and in need of protection. Thirdly, Company T has a different attitude towards R&D.

Company T spends 5% to 10% of turnover on R&D, which is a very high proportion for the machine tools industry. The R&D function is centralised except for a new branch involved in robotics.

The Company T respondent perceived the company as fairly innovative but considered that most radical innovations come from small companies set up to develop an idea and not from large companies. Moreover, Company T is in a mature industry and is, therefore, more concerned with progressive iterations rather than radical new innovations.

The technologies on which the company bases its innovation are electronics, computer technology and polymer science as a basis for plastics research. Company T perceives that there are three areas which will be increasingly important in the future—robotics, materials substitution and biotechnology. Company T has research projects in the first two areas but has decided not to do research into the third. Company T decided to go into robotics 10 years ago and it took 5-7 years to consolidate its corporate thinking behind the effort. Over the last five years it has implemented the research and set up a separate operating division for robotics. This new division was created to allow robotics to grow quickly. Development was separated from research so that new products could be brought on rapidly.

Other innovations have taken less time; for example, entry into flexible manufacturing system (FMS) technology took three years plus some time for secondary development. The time taken to get a new project off the ground in Company T depends on the amount the innovation departs from existing products and on the complexity of the job. Developing for new markets takes longer or has to be done in a smaller way.

The Company T respondent suggested that once an innovation has been adopted, several factors influence the speed of progress. These are the size of the market (the company aims to gain 50% of the market in the first instance), the ability of the company to provide the right environment for the innovators, and the venture capital available within the company. Innovations are not made in a vacuum but rather aim to fullfil a need. The specifications for innovations arise from these requirements but may be misjudged at first so there must be flexibility to change the specifications as the project progresses. Too narrow restriction is a barrier if it occurs. Company T sees itself as in a mature industry where market demands are already set and are quite rigid. The requirements for new products are, therefore, normally quite specific.

Secrecy was regarded as reasonably important in the sector but companies tend to know what others are doing, especially in the process rather than product area. Patents were seen as important as a bargaining counter.

With most new products Company T makes very thorough market assessments on the potential market, the competition and so on. With robotics, market forecasts were more difficult to make, and the more different the innovation, the more this is the case. For Company T, sensitivity to the market share that it can hope to obtain is vital in making decisions to innovate. Company T is continually looking for new niches and for more product lines. This broadening of its product base is driven by the desire to increase its market share.

The Company T respondent did not believe that research can be costed tightly. In making cost assessments for innovations the company tries to make forecasts and predict the effects new technologies have on these. Actions of competitors were seen as important in the innovation process because they affect the market. Developments which occur in other industries, such as electronics, must be monitored in case they offer advantages to the machine tool industry.

The Company T respondent suggested that not one but two champions were needed for innovation. The first must be close to the technology, and is the person who has the idea which he then pursues and sells to the second person. The second person must be in senior management and must take up the idea and persuade the company to support it. It is rare, he suggested, to find creative technologists who also have management capabilities.

The Company T respondent felt that the most important factor contributing to innovation in a company is management attitude and both senior and junior management must support it.

IV. Sunrise Industries

THE BIOTECHNOLOGY INDUSTRY

Overview

Biotechnology has been hailed as the catalyst of a new age of development, promising spectacular opportunities throughout many industrial sectors. It is seen by many academics and industrialists as being one of the major new technologies capable of having impacts throughout industry, perhaps in a way similar to microelectronics. The rewards of early investment have sometimes taken time to materialise.

Biotechnology, defined as the application of biological organisms, systems or processes to manufacturing and service industries[53], may be seen to encompass a range of distinct technologies. These include: genetic engineering (or recombinant DNA); enzymes and enzyme systems; monoclonal antibodies; and immunoglobulins; plant cell culture or single cell protein and interferons. Hence biotechnology promises applications in a wide range of industrial sectors from food and drink, to chemicals, pharamceuticals and agriculture and offers opportunities to find new sources of energy and pollution control.

This very diversity and broad spread of applications causes problems in coordinating research on a national basis and the UK effort has been criticised for lacking coherence and urgency [54,55].

---

53 ACARD, ABRC, RS Biotechnology. Report of a Joint Working Party, HMSO, London, March 1980 (The Skinks Report).

54 Consultative Board of the Joint Consultative Organisation for Research and Development in Agriculture and Food, *Towards a Policy for Research and Development in Agriculture and Food*, JCO Consultative Board, June, 1982.

55 For further reading in the field of biotechnology see the following: The Technical Change Centre itself has embarked upon a major study of the affects of biotechnology on the food industry. Bull, A. T., Holt, G., and Lilly, M. D., *Biotechnology International Trends and Perspectives*, OECD, Paris, 1982.*Impacts of Applied Genetics, Micro-organisms, Plants and Animals*, Office of Technology Assessment, US Congress, Washington D.C., April, 1981. Holmes, A. W. and Jarris, B. *Application of Biotechnology in the Food Industry — An Appraisal*, Leatherhead Food R.A., January, 1981.

The business community in the USA has been keen to invest in biotechnology. It is already a growth business with hundreds of small firms dealing in biotechnology. US government funding for biotechnology is generally thought to be less important than in Europe, except in the field of energy, and to a lesser extent enzyme technology, because of the availability of venture capital. There is substantial activity in parts of the pharmaceutical and petrochemical industries, and ample capital has been poured into aspects of biotechnology, demonstrating a general confidence in the potential of biotechnology to the US business community.

The Spinks' report compared activity in the field of biotechnology in the UK, the USA, France, West Germany and Japan and found that the level of awareness and the sense of urgency shown by overseas firms was lacking in Britain. Investment in biotechnology in the UK had been slow and opportunities had been missed.

Much of the early practical support for the development of biotechnology in the UK came from giant companies such as Tate and Lyle, ICI, Rank Hovis McDougall and Unilever which embarked on extensive research projects in the sixties. Some are now beginning to see marketable results. City finance houses in the UK now seem ready to invest, and funds are emerging from larger firms set up for the purpose. Since the UK business community was slow to react to opportunities, a new phenomenon has emerged. Groups of small firms have been established to exploit developments in biotechnology at specific centres such as Plant Science Ltd., linked to the Wolfson Institute, University of Sheffield, and Imperial Biotechnology Ltd., linked to Imperial College, University of London. This reluctance to seize biotechnological opportunities may be a result of the long lead times for research before commercial products are available, and the huge costs of development.

The controls and constraints on recombinant DNA research in the UK are more severe than in some other countries. It has been claimed that present legislation in the UK, which requires the reporting of all research proposals in the genetic manipulation field to the Genetic Manipulation Advisory Group (GMAG), jeopardises future patents and renders the UK unattractive for research investment[56]. By contrast in the USA there have been substantial relaxations in the guidelines provided by the National Institutes of Health (NIH).

56 Op.cit.

Company I

Company I's biotechnology work is based at its main R&D laboratory which is the largest of several laboratories owned by this major British company. There are 1,250 people employed on site.

The company has a lead in large-scale application of tissue culture propagation in which it has had an interest for over 10 years. The US interest in this area is more recent and tends to be applied to house plants.

The Company I respondent stated that in Britain the biotechnology industry is largely dominated by big firms, for example Unilever, ICI, Glaxo and Burroughs Wellcome, and he suggested a need for growth of small supplying and ancillary firms. He partly explained this different pattern of development and the lack of growth in these small firms to be due to an extent to the lack of competent scientists to run them.

Company I is concerned with many technologies. The major development, which is already in use, is based on propagation of clones by tissue culture. This is used to multiply the best specimens for widely used tropical crops.This technique could increase average yield by up to 30% and produce a product more suited to certain applications. The use of enzymes is being researched both for catalysts in production processes, and for producing specialist products. Fermentation technology using micro-organisms and concentrating on high margin products such as perfumery chemicals and flavourings is also being considered. Other research is based on the plant technologies of recombinant DNA and animal culture, biologically based sensors and fundamental cell biology. Company I felt that production of biomass by fermentation is uneconomic in comparison with agriculture whatever its end use.

The Company I respondent felt that although Britain led the world in fundamental research in biotechnology in the 1950s, it is now second to the USA. The speed of innovation, it was suggested, is hampered by universities in both Britain and the USA where subjects such as fermentation are not given enough attention. It was suggested, as in Company J, that there is inadequate interest in plants as opposed to animals. Work on recombinant DNA in plants is lacking and there is an over emphasis on applications for the pharmaceutical industry. Although Britain has an Agricultural Research Council, it was felt to be performing unsatisfactorily in this area.

The lack of work on screening micro-organisms in the UK and USA compared to Japan, and the high costs of research also act to slow down progress. Research is costly because it has to spread over growing seasons, it requires many support staff, and uses high cost chemical materials. In addition, the respondent suggested British industry is weak at finding out what is going on. It is too secretive, and lacks enough of the strong internationally respected research groups which function to bring together information from all over the world. This combined with the lack of co-ordination in government funded research hampers the speed of innovation. Innovation in Company I was also affected by the general economic climate and although the company can attract good scientists in adequate numbers, it has problems in recruiting ancillary staff.

Although Company I limits its work to plant science areas, it is aware of the large US effort in diagnostics and that the US industry has a possible winner in methods of interferon production. The respondent noted that in the USA, pharmaceutical companies have been buying up seed companies, although the exact line of work that they had in mind was not known. The Japanese were regarded as strong competitors on fermentation, and were extending its range by genetic manipulation. They are also strong in the development of sensors for diagnostic use.

Company J

Company J is a US consultant in biotechnology with particular reference to the seed and plant areas. The company is small, and advises on mergers, acquisitions and strategic planning. The company does no basic R&D but works at the interface of seed and plant science which is seen as the delivery system for the technology.

Company J is concerned with technologies such as recombinant DNA, plant tissue culture, plant breeding and hybridization agents. Although the most important technology for it and others like it is still basic plant breeding, the most dramatic technology is the hybrid technology. For example, heterosis can give a 10% to 15% yield advantage, offers protection in that the parents of the plant cannot be identified from its appearance, and provides an annual business.

Decisions to proceed with innovations in biotechnology are generally perceived as offensive rather than defensive; that is, they are positive diversifications to take advantage of new opportunities. The only defensive entries into biotechnology have been in the drug industry, where companies have moved into gene splicing. Reasons for acquiring seed and plant companies were primarily commercial and economic in the 1960s, but advances in the manipulation of bacteria provided technological reasons for acquisitions. New plant genetics dramatically enhanced the productivity of plants and with modified plants it became necessary to control the delivery system, that is, the seed companies. The Company J personnel involved in innovation in this area have PhDs with two to five years post-doctoral experience and work experience.

The company is involved in the entry of other companies into biotechnology, by performing external assessments of the opportunities for them. The initial assessment always involves four questions—What is the market? What is the technology? When is the market going to be revolutionised? Who are the competition?

Cost assessments are made with reference to market size. Return to research investment is calculated for the worst, best, and reasonable cases and is costed in terms of the professional and technical staff required. The time necessary to meet objectives is set, and costs calculated—10% to 15% is added to this sum. The most difficult costs to calculate are development costs, which are also the most expensive and critical since every dollar error reduces profits. This process takes six months to a year. Recommendations are passed onto the board, and a decision is made either to do research or to acquire an operating company through which efforts can be channeled. By the time the money is available one to two and a half years have elapsed. It then takes another year to put decisions into operation.

Once companies move into the technology, it is not the level but the length and uncertainty of funding that is problematic. Innovation takes time and a steady guaranteed amount of funding is necessary. At least five years are needed for agricultural projects. Annual budgets can be unsettling and disruptive. Another source of delay is the shortage of good people in the plant science areas. PhDs tend to move into the bio-medical area and gaps in the plant science field are not being filled by those committed to biotechnology, but by those who have been retrained. "People with a good long track record are very few". The Company J respondent felt that in the USA agriculture suffers from a lack of interest by government, and cited the lack of a national centre for agricultural research as an example.

For Company J, research in the biotechnology field must be flexible. Companies may have a general headline area in which they work but the real benefits arise from unplanned spin-offs. The companies which have suffered the biggest disappointments in the USA are those which have stuck too rigidly to their original goal.

Innovation in biotechnology is slowed by the inadequate supply of qualified scientists, and by the competition between biotechnology and other new technologies and therefore a lack of thoroughness or commitment to biotechnology. Other factors seen as important in Company J were the general economic climate, actions of competitors, government intervention through tax concessions and, most importantly, difficulties in achieving sustained finance.

The Company J respondent pointed to the major differences between venture capital availability in the USA and the UK. In the USA this provides the individual with funds to follow ideas, to research and to attract good people.

In the field of agricultural high technology, firms have been successful in attracting talent. They pay well, and have a management structure which is not too rigid. In contrast, firms in agricultural genetics have not been successful in producing products because of the time involved and the effects of public funds which are available in this area. Small firms in agrigenetics get little revenue from their products. Agricultural biotechnology companies were regared as more successful.

A product champion was considered important to successful innovation in Company J. He could be on the board, in middle management, or in R&D. The position of the person in the company is not as important as personal qualities such as self-motivation and single mindedness. An innovative company must show commitment to innovation from the top and have the faith to do R&D, and to look for options. It must employ good scientists and pay them well. Finance is no problem in the short-term, but must be committed over long periods with reaffirmation of targets rather than threats to withdraw finance.

THE MEDICAL IMAGING INDUSTRY

Overview

For years medical imaging has been dependent on radiology or x-rays but recently there has been a major technological advance in this field, i.e. the development of whole-body scanners based on the nuclear magnetic resonance (NMR) principle. NMR is a physical phenomenon which has been used for some years in the qualitative and quantitive analysis of large organic molecules, including those which occur in living tissue. Industrial applications have continued to increase.

In the last 25 years NMR spectrometers have been used widely for research in chemistry, and in the drug and food industries. A NMR spectrometer generates the pulses of radio waves, detects and records the resonances, and helps to interpret the results in terms of chemical and structural composition. The ability to resolve fine molecular detail is much increased by using high magnetic fields which can in practice be provided only by superconducting magnets. These form an integral component of most modern NMR spectrometers and account for 10% to 30% of the price. One of the cases presented in this section is of the leading manufacturer of superconducting magnets. The other case is a company involved in the manufacture of the complete NMR whole-body scanner. The first company is a supplier to the second.

There are now some 30 NMR whole-body scanners installed in hospitals, mainly in the USA and Europe. The estimate for 1984 is 160, over five times more. The idea was developed at universities in the USA and UK but the superconducting magnets used in NMR whole-body scanners were developed by a company in the UK. Development work is still in progress in hospitals, and in the laboratories of the principal systems and component manufacturers. Forecasts of the eventual market vary, but current interest indicates a growing demand. NMR is proving to have many advantages over other imaging techniques, notably x-ray computer assisted tomography (CT) scanning. NMR gives good pictorial images, based on water content and other chemical differences between tissues and organs, which can be combined to emphasise particular information. As images can be obtained in any plane by varying the procedure, three-dimensional information is readily available. The information is processed by a computer and plotted on a TV screen, producing an image of the underlying anatomical shapes with some indication of what the tissue, normal or otherwise, is composed. In addition, NMR can see past bones, which x-ray CT cannot do, and good images can be obtained from soft tissues close to heavy bone structures. This is particularly useful in the case of the brain, spinal cord, knee, eye and pelvic region. NMR can detect the early stages of multiple sclerosis, and show growing brain tumours which CT can miss. The patient avoids exposure to x-rays or other harmful ionising radiation, and generally there is no need for injection with a contrast medium. There are no known harmful side effects from the exposure of the human body to the magnetic fields and radio-frequency waves in this equipment. NMR can also give more useful biochemical information than is available from tissue removed surgically in a biopsy, which may be chemically altered by the fact of its removal from the living body.

Outside the central nervous system, the advantages of NMR over CT are not so clear, and CT scanners have two major advantages: they are time tested, and less expensive. The question then is how much more performance can NMR offer, since it can cost twice as much as a CT scanner, and cannot be used on patients with pacemakers or respirators. Manufacturers of NMR whole-body scanners must obtain the approval of the US Food and Drug Administration (FDA) prior to commercial marketing in the USA. The Radiological Devices Advisory Panel of the FDA has recommended to the FDA that the devices supplied by at least three of the manufacturers be approved subject to certain minor conditions such as labelling revisions. Final approval is anticipated within the next few months.

There are about 10 major companies involved in NMR imaging development throughout the world. It is likely that those companies which are expected to have large shares in the market for NMR whole-body scanners will manufacture some of their own magnet requirements. At present, however, there is one leading manufacturer of the NMR imaging magnet based in the UK. There are also a number of smaller companies offering superconducting magnet systems in other specialised areas. Although a number of potential industrial applications for superconducting magnets have been recognised for some years, various practical obstacles, principally connected with the need to operate them at low temperatures, have delayed progress.

The markets for both the magnets and the imaging machines themselves are in Europe, Japan and the USA. There are some manufacturers located in each of these regions. Successful UK exporting firms are faced with a strong pound and inflation rates which are higher, and in the past have been much higher, than in many countries to which they are exporting (particularly in West Germany, Japan and the USA). To offset this, retain competitiveness, and accrue satisfactory profit margins, it is necessary to obtain sharp reductions in unit cost, through improved efficiency or productivity, or to gain much higher sales. Continuing success is dependent on products technically in advance of the competition.

Company K

The Company K group is based in the UK and is the leading manufacturer of a broad range of superconducting magnets. The group which started in 1959, was born in a university research department when extramural services to academic laboratories led to larger orders from government establishments. Regular contact is still maintained with 12 universities around the world and the origin of many of the group products are from customers in these universities.

The first significant order for Company K was for two large technically advanced but conventional magnets, which were made in a garden shed. At this time sophisticated electronic power and water-cooling devices were required for laboratory work in the high magnetic field area and the market was very restricted. After a conference in the USA the founder of Company K decided to manufacture superconducting magnets which did not require these power plant and water-cooling facilities. Not only would the market have vast potential but the development involved low capital outlay to produce compact low-weight magnets. The only significant problems were dependence on external sources for an adequate supply of liquid helium and of suitable material for magnets. Both proved to be lasting problems and the former led to the setting up of a separate company investigating cryogenics.

During the early years there were considerable technical problems which were somewhat eased by the production of an improved niobium—titanium wire by another British company. The flow of significant scientific and technological achievements continued, and enabled Company K to find finance. This masked the fact that development of management and commercial skills in Company K did not match technological achievements. A further financial crisis heralded a change in management, and from 1973 profits grew rapidly, particularly in the field of NMR spectrometry. Diversification has also been actively pursued mainly in the field of medical instrumentation. Company K's policy is to set up a separate company for each new product.

The group now consists of seven semi-autonomous operating companies, each catering for a different product area, and reponsible, with clearly defined limits, for its own affairs, helped in terms of broad strategy and finance by the group. The group has two sales subsidiaries in the USA, one in West Germany, one in France, and two smaller ones in the Netherlands and Switzerland.

The group entered the 1980s in a period of growth. From 1978-1979 alone turnover had increased by 57% and exports by 71%. The basic elements of the group's strategy are scientific excellence, an emphasis on technically advanced products, and a concentration of effort in fields where they have special expertise in order to maximise the likelihood of keeping ahead of competition. The company is continually alive to the possibility of new applications of superconducting magnets.

Company K see recruiting the best brains and keeping them as a primary objective. The average number of employees rose from 500 in 1979 to 950 in 1983. Of this 950 over 200 are overseas, 135 have degrees of which 34 are doctorates. In the year ended 1983 approximately 10% of turnover was spent on R&D, spread throughout operating companies. Two-thirds of the group's revenues came from exports, the biggest customers being in the USA.

There are three technologies essential to the superconducting magnets with which the company is concerned. These are superconducting materials technology; cryogencies—that is, the technology of keeping things cold (within a few degrees of absolute zero); and high magnetic fields, in which Oxford University, which they are near, is one of the world centres of excellence.

Many new ideas came from Company K's customers (particulary universities). Decisions to proceed with innovations will always depend on perceived market demand which must be coupled with cost estimates, profit forecasts and consideration of the competition. Innovation decisions are taken rapidly since the likelihood of a high-tech product retaining its timeliness is not high if delayed for long.

The latest example of innovation in Company K is in the field of NMR imaging. The aim was to increase the magnet from 30cm to whole-body size. The decision to proceed with the programme was taken in 1979 and the first prototype magnets were delivered within two years. By 1983, with 31 deliveries, the project was reaching maturity and a separate company was set up to take over this part of the business. This process was consistent with group policy to form a separate operating company for each established new product group. The speed of progress in this innovation was affected by the fact that although well understood technologies were used, they were stretched beyond previous limits. Unexplored constraints were placed on the properties of materials used and once these had been mastered the normal "learning curve" applied.

Initially the latitude in design in this innovation was large since any whole-body size superconducting magnet had a guaranteed market. As the project progressed, within the company, specifications tightened. The early decision to proceed was difficult because of the uncertainty of achieving the technological goal, and the adverse financial consequences of failure.

Company K, in assessing the market for the new product, consulted virtually all the world's major health care suppliers (the primary market). Market planning flexibility in Company K results from plants operating in both the UK and USA giving a best and worst prediction of market share which varies by a factor of three. In this case even the worst prediction offered a considerable business opportunity.

Cost assessments of this innovation in Company K were made by scaling up from well established small-scale products. Uncertainties lay in the lead time expected before selling the first products (in the event, shorter than anticipated), in write-off of prototypes, and costs of required materials, in particular liquid helium.

Secrecy is regarded as vital in Company K when developing a new innovation. For the company non-mobility of key staff has been an important factor in preserving secrecy and guaranteeing success. The Company K personnel involved in the innovation are all scientists or technologists who have at least five years of hard business experience and have risen to decision-making positions. Many have PhD's, and most have been with the group for more than 10 years and so have detailed knowledge of the specific markets in which it operates. This innovation in Company K was championed by the managing director of the subsidiary company involved in the development programme.

Highly relevant to the innovation process in Company K is the US health care market's ability to absorb sophisticated, expensive medical instrumentation and the success of the previous generation of x-ray CT scanners which both paved the way and showed likely pitfalls to be avoided.

Company L

Company L is one of the three or four US companies which produce the NMR whole-body scanner. It, like Company K, has increased turnover rapidly since 1980, doubling it annually. The company sells directly to the medical profession, and manufactures a range of new diagnostic devices including ultrasound systems and mobile x-ray.

Company L was founded in 1977 to make ultrasound diagnostic systems to fill a gap in the market perceived by its founder. At the top end of the range machines provided high quality still pictures. At the lower end, non-computerized models showed movement but their poor quality images could not be stored for further study. The company developed a middle range product combining the best of previous ranges. Ready acceptance was achieved and others followed rapidly. Although smaller than many of its competitors, the company was able to keep its prices low to nurture US sales while solidifying its reputation as a technological leader. Other medical appliances based on ultrasound technology served to keep sales growing in this area.

The company soon diversified into DSA (Digital Substraction Angiography), like CT scanners a marriage of computers and x-rays, but much less complex, and within a year DSA accounted for 26% of company sales. This was only the first of what the industry expected to be a generation of digital radiography products and Company L, with its early entry into the market and well designed product, managed to gain a 20% market share. Since the company visit, Company L has written off this whole sector of the business with a loss of $60 million. The digital radiography market did not develop to the size anticipated and was no longer financially viable as a stand-alone business.

Working on a similar principle to Company K, Company L has set up separate operating units for each of its products. Its latest unit concerned with NMR delivered its first NMR whole-body scanner in 1983 and industry estimates suggest that the total market will grow from 50 or 60 units in 1984 to 500 to 700 by 1987. To enter NMR, in 1981 Company L bought the patent rights and a prototype owned by another company and took over the University of California, San Francisco, research programme that had been sponsored by the latter. Company L follows the market pull theory of innovation working on the principle that the best technology in the world does not necessarily give advantage unless one has strong sales and marketing efforts.

Company L has a small central R&D function but most R&D is done in the divisions. R&D is also contracted out to universities or research organisations. Company L has a foundation of computer based technology and intends to continue to innovate in that area.

Company L regards itself unquestionably as an innovative company. Innovation is a way of life which allows a small company, by maintaining its technical edge, to keep ahead of many of its larger competitors. For Company L, another important factor in deciding whether or not to proceed with an innovation is clinical relevance. The company listens to its customers and takes its direction from clinicians. Decisions to go ahead with an innovation are made quickly.

The company is growing so rapidly that finding good personnel requires effort. Company L is working on the edge of the technology and needs the latest electronic devices so the supply of key items for innovation can also cause problems.

The latitude in design of new products in Company L is considered reasonable but is limited by cost, and directed by the market. "It is imperative to bring something new to the market-place all the time, even if it means holding back on something the first time round."

Costs of innovation are calculated in terms of return on investment within the market constraints. A certain return is needed before a project can go ahead and each product must stand alone. Occasionally Company L decides to go ahead with an innovation because it provides some benefit other than a new product. The unpredictability in Company L's assessments lies in forecasting market size.

In Company L, secrecy is important in the development of an innovation but it can never be complete in the industry because work is not conducted in a vacuum. Customers must be informed and kept up to date with developments and this inevitably will filter through to the competition.

The senior personnel involved in the innovation process in Company L nearly all have PhDs in technical fields. The management all understand the technologies they are involved with and this is felt to be very important as is a champion for innovation.

For Company L the most important factor in being innovative is the quality of personnel. It is necessary to have good people who are hardworking and committed, and this commitment to innovation must come from the top of the company. Company L's ability to sustain the major loss that it did in 1983 illustrates its strength in most areas.

THE ELECTRICAL/ELECTRONICS INDUSTRY

Overview

The evolution of electronic technology, microelectronics, has, over the past decade, been very rapid. There have been steady, quantitive improvements in the technology, with smaller and smaller electronic components performing increasingly complex electronic functions, at ever higher speeds, and at lower cost. These changes have meant a qualitative change in technology. Many of the technological achievements of the past decade have depended on microelectronics, and microelectronic devices are the essence of many new products from communication satellites to digital watches. The major impact of this new technology has been to make electronic functions more reproducible, more reliable and much cheaper.

With technical developments, costs have decreased and a widening range of applications have become possible. The development began with the transistor and was almost simultaneous with the development of the stored programme digital computer which provided a large market for transistors. Thus, a new component and a new application gave rise to very rapid growth. From the transistor were developed the solid state integrated circuits.

Early efforts to miniaturise electronic components were not motivated by computing needs, but by defence needs for complex electronic systems in which size and weight restrictions were very severe.

The eventual solution to problems of minaturisation was provided by the semiconductor, integrated circuit. The integrated circuit developed at Fairchild in 1959, accomplished the separation and inter-connection of transistors and other circuit elements, electrically rather than physically. Additional techniques have been devised, giving the designer of integrated circuits more flexibility, but the basic methods were available by 1960.

Development has been very rapid and an individual integrated circuit on a chip, perhaps a quarter-of-an-inch square, can now embrace more electronic elements than the most complex piece of electronic equipment that could be built in 1950. Experts consider that even though the growth until now has been so immense, the technology is still far from the fundamental limits imposed by the laws of physics and that further miniaturisation is less likely to be limited by the laws of physics than by the laws of economics.

The growth of the microelectronics industry shows the result of investment in research followed through with entrepreneurial flair. After the introduction of the integrated circuit in the early 1960s, total world consumption of integrated circuits rose rapidly reaching a value of $1 billion (US) in 1970. By 1976, consumption had more than tripled to $3.5 billion. The impact on the electronics industry has been immense; in electronic equipment, less than 10% of the value is in integrated circuits, but most of the world's $80 billion electronics industry depends in some way on them.

The substitution of microelectronic devices for discreet components reduces costs because the devices are cheaper, and the integrated circuit contains many of the inter-connections previously required, hence, it saves labour and materials. The inter-connections of the integrated circuit are more reliable, and this leads to savings in maintenance. Integrated circuits are smaller and consume less power than the components they have displaced, they make savings possible in support structures. Less testing is needed in the course of production because the correct functioning of the complex integrated circuit has already been assured. Eventually, the end-user needs to provide less floor space, less operating power and less air-conditioning for the equipment.

The most striking characteristic of the micro-electronics industry has been a persistent and rapid decline in the cost of a given electronic function. Part of this decline can be accounted for in terms of a "learning curve". Because of the rapid growth in this young industry, cost reductions have come at a much more rapid pace than in a mature industry. The cost of a given electronic function has been declining even more than the cost of integrated circuits since the complexity of the integrated circuits has been increasing as their price decreased. These cost declines were accomplished by the traditional learning process and by the integration of more bits into each integrated circuit.

The microelectronics industry has served a proliferating market of ever-broadening applications. As each new application consumed more microelectronic devices, more experience has been gained, leading to further cost reductions and opening up a wider market for the industry.

Cost reduction has resulted from the development of increasingly complex circuits that lower the cost per function for both the circuit producer and the equipment manufacturer. The main technical barrier to achieving more functions per circuit is production yield. When the cost of scrapping exceeds the cost savings in subsequent assembly and test operations, the cost per function increases rather than decreases. The most cost efficient design is a compromise between higher assembly costs and higher scrapping costs. Technological developments have concentrated primarily on increasing the production yield, either by reducing the density of defects or by reducing dimensions. Attention in process control to cleanliness reduces the defect density, and improvements in photo-engraving processes reduce the dimension of the basic circuit elements. Electron beams or x-rays are being developed to reduce the dimensions even further. The circuit becomes faster as it becomes smaller. "In an industry whose products decline in price by 25% a year, the motivation for doing research and development is clearly high. A year's advantage in introducing a new product or new process can give a company a 25% cost advantage over competing companies. Conversely a year's lag puts a company at a significant disadvantage with respect to its competitors. Product development must be a critical part of company strategy and product obsolescence must be accepted as a fact of life[57]."

57 Noyce, R. N., "Micro-electronics" in *The Micro-electronics Revolution*, Ed., Forester, T. Basil, Blackwells, Oxford, 1980.

Company M

Company M, formerly a service group within a large UK company, is now an automonous subsidiary formed in 1980. It has developed a series of advanced CMOS, Uncommitted Logic Arrays, and a design technique based on the use of functional cells modelled on 4,000 series parts. In addition to ULAs, Company M provides a complete custom LSI service for users requiring devices in medium or low volumes. The service includes full design, assembly and test facilities. Company M also operates as an indepentent Assembly/Test House. Company M is the biggest European design centre independent of a manufacturer.

The Company M respondent described the business environment in the semiconductor industry as aggressive and innovative, fast changing both in terms of business technology and business methods. The scenario, he suggested, is very different from that found in other industries. The customer environment dealt with elsewhere is quite different and much less aggressive. He regarded the semiconductor business as a small club in which everybody knows those with whom they deal, and their reputations. This changes the nature of business relationships. On the whole people in the business, he felt, were entrepreneurial and enthusiastic, and in this particular business world it would not be possible to succeed without innovation. Company M sees itself as having a technology looking for an application; it is not responding to customer needs. The market in which Company M operates can be divided into majors, intermediates, and minors. The Company is at the moment minor but it is hoping to move into the intermediate group. It would not be possible for it ever to be a major.

Company M's innovation concerns uncommitted logic arrays. The company has many new ideas associated with products and with design. Its innovations are different from those which arise in the normal R&D environment. The innovation need is for a product consistent with an overall strategy, from which no digression is permitted and Company M's work is all channelled to this end. Company M has now developed a new family of uncommitted Gate Arrays, and two more are being produced for particular new needs. Innovations in Company M are difficult to describe in once and for all terms. The company is working with a very short product life cycle of two or three years, and is under continuous pressure to innovate.

The original innovation developed in response to a realization of a growing market need for a fast response customer design service.

Fundamental to the business strategy of the parent company was a role for customer design, involving high technology for its own products; thus this function was identified and separated out. The idea was developed and then it was offered to other customers, in a custom and semi-custom range. The development went through a process of satisfying an internal need for the parent company and then it was taken forward by three particluar people. The prime factor in the innovation process was the identification of market need. There was some quantification of likely costs but more as a rationalization. The real moving factor was an experience based value judgement " . . . of course if you decide in this way sometimes you're right and sometimes you're wrong. But having taken the decision, the project group within the company decided to proceed whatever happened". In this innovation in Company M personality factors dominated. Clearly the effectiveness of the personalities involved was dependent on their credibility rating within the company, which related to their success in the past in developing new innovations. For Company M there are two aspects of the push for a particular innovation: one is the technology and one is the commercial push, and both were strong on this development. The decision and innovation process took about 18 months to two years from conception to product launch. The first year was within the parent company and the second year was in Company M. There has been a remarkably large improvement in creativity since Company M broke away from the parent. Company M is now in a battle for competitive success, a very new environment, and without the protection of the parent company. It now has a new product with a product development time of 12 months. The time frame for creation and development is getting shorter.

For the Company M respondent, progress of innovation in this area can not be speeded by applying more people or more skills to a particular problem. The process does not work this way. The aim in Company M is to support the few creative people with original ideas, and involve them throughout. Innovation, must be adequately supported, otherwise the people with ideas become disenchanted. A management team who can identify possible innovations, respond to and support them is vital.

Since an innovation is a new idea, this management group in Company M needs to be close to the innovation process, but there is a constant risk of complete failure and they have to work at great speed: that is, at the speed of their initial reaction. In these circumstances it is not felt possible to have a formal well worked out assessment. Company M engineers need to see their ideas go through if they are to go on being creative. It was suggested that this speed of reaction time is probably industry specific since developments are very fast in the semiconductor industry. Finance, can be a problem in a small company because every risk is a big one, though it might not seem so to big competitors.

In Company M progress in the industry is crucially affected by good aggressive marketing. Company M does not keep ideas to itself but licences them all over the world, and this has created customers, multi-sourcing arrangements, and so on. The Company M approach was contrasted with that of a major British electronics company which was at the same point in terms of technological development in this area but had decided to be more secretive. This, it was suggested, was a typically British strategy which failed in that although the project was not a complete failure, it was not the success that it might have been.

Secrecy was not regarded in Company M as important to successful innovation. Success depends on doing the job properly, and over attention to secrecy was regarded as a British mistake. Within the semiconductor industry, the Company M respondent felt that people solicit ideas, which are freely exchanged in the right circumstances. Cooperative ventures are very productive. There might be a limited degree of secrecy before licencing, but it is impossible to keep secrets forever and it is better to benefit from the licence.

Company M regards the Gate Array business as highly competitive with many new contenders, so success requires technical and commercial credibility. The speed of progress of an innovation in Company M can be affected by difficulty in finding the solution to technical problems. Company M has had to slow down at some stages in projects to solve particular problems which were not envisaged at the start. Encountering new problems in this way could be a major stumbling block unless the management team are completely committed to innovation. In Company M this commitment to innovation is further encouraged by training the company accountant in marketing and in production, so that a dimension is added to accounting judgements on the risks that are to be taken.

The Company M respondent felt that the parent company had shown great entrepreneurial flair for a British company in allowing such developments within their broader more traditional frame work.

The break with the parent was, for the Company M respondent, of particular significance since Gate Arrays in the USA almost always come from small companies which suggested an optimum size for innovation in this field of probably less than 500. In America, he noted, this keeping within size limits often happens by default when two or three individuals unable to push their idea through in a big company leave and start up on their own. It is now so well recognised in the US that companies are planning such patterns.

The technologies essential to the innovation in Company M are of CMOS in the digital area and in the linear bipolar area. These technologies are mostly non-UK technologies. The Company M respondent felt they might have been British technologies if the relationships with British companies had been more positive.

Company M feels that business strategy should always involve setting up a new business with its own accountability if a developing new idea should merit it. Given these views about size and accountability, Company M was always more inclined to deal with smaller firms rather than large ones, since Company M personnel could have more direct contact with the people with whom they would be discussing technical problems. Since Company M does not make silicon, it is considering packages supplied by the Japanese, regarded as very good suppliers.

In Company M the progress of innovation is affected by the fact that so many unanticipated ideas are being generated that the choice of which is likely to succeed is very difficult. Another of the problems in innovation in microelectronics is the response time which can be crucial since developments are iterative.

Company M allowed total latitude in the original thinking about the innovation, then it moved towards very specific constraints on development to ensure a product acceptable to the market-place. The guidelines were laid by four people, two engineers and two managers, who took a collective "Japanese style decision". After this there was a policy of no deviation, unless there was a very pressing reason.

The original advice from the company's researchers was that the project sounded good, but that prediction was uncertain in a completely new market. The company analysed the market strategies of competitors, particularly British competitors, and decided to opt for an American approach which seemed likely to succeed. It felt no exact quantification of the market was possible, since there was only a very general outline available.

Customer product strategy was already established in the parent company, which had a flexible marketing plan involving pursuit of its aims but with a willingness to change tactics as outside circumstances changed. The strategy remains unchanged but new tactics may be adopted; thus the company has changed from a totally customised product to a semi-custom one.

Once the original innovatory idea was accepted, financial assessments were then produced. These concentrated on manpower and attendant costs. This assessment proved inaccurate. The project cost more but it was decided to continue in spite of this. "Originally we made our assessment and doubled it, and in fact the project cost twice as much again. The major point of uncertainty involves how long the project will take, and how much labour it will involve, and also the number of iterations for success—that is how many times it will be necessary to go around the development cycle. Manufacturing iterations of an innovation cost a great deal of money."

The people involved in the initial innovation decision in Company M were an engineer who had the basic idea, which was then developed with two others in engineering management and one in marketing. The first engineer pushed the idea hard. His audience was receptive and the idea proceeded slightly modified. This all happened within the parent company's Profit and Loss Centre (PL Centre), an autonomous management centre. That it was autonomous was thought very significant since it meant that a small group of people could accept or reject the idea. It was suggested that the outcome might have been very different if the idea had arisen in a big team which was part of a big bureaucracy. Originally the parent company had a Design Development Centre in their Engineering Department, and from this they set up the semiconductor P and L Centre. This was regarded as a popular strategy within the industry.

The senior personnel involved in the decision were mostly graduate engineers with post-graduate experience. The key management team within Company M now are all engineers with management training. The marketing manager came originally from an equipment manufacturing company, and is also an engineer.

A major champion for the innovation was regarded as very important in the formation of Company M. Originally there was a champion both at the operations management level and at board level. It was thought very difficult to get ideas off the ground without support on the board. "Often the understanding of the board is not highly technical and therefore someone who can put ideas across to them is necessary."

The general financial position of the company was regarded as obviously significant in innovation, although there had been no specific problems of finance for this scheme. The Company M respondent felt that had this innovation been starting cold, not within the confines of the parent company, then the problems of getting risk capital in the UK would have been very great, although it would have been possible to go to the US where venture capital is easier to find. In fact the company got significant government support through the Department of Industry and feel that government, in this case, has helped them a great deal.

For Company M the supply of qualified personnel of the right calibre is always a problem in the industry, but it was not a factor in this particular case. The move to a new company and away from the parent company was seen as helpful in Company M, but the company had a problem of recruiting a manager experienced in production which led to a change of personnel and this tuned up the management team.

The influence of Company M customers on innovation was seen as very important. Innovation is highly directed to the market and there is a constant need to fight new competition. The actions of existing competitors also influence and stimulate developments. Company M is now a pioneer in the field.

The importance to innovation of the movement of ideas was stressed. It was suggested that in Japan cooperative management efforts within large corporations ensure this. In the US movement of personnel with ideas between companies has the same result. In the UK none of these things occur so ideas, although developed, and often very good, stay in certain places, do not move about, and do not get developed any further. It was further suggested that it is possible for governments to finance their way out of this kind of situation, as was happening in France, but that the British Government was not tackling this problem.

Company N

Company N's sales have grown from $4.4 millions (US) last year to approaching $10 millions this year, and the company is aiming for $25 millions when its new product comes on stream.

It employs 105 people at four sites arranged according to function. The Chairman of the Board, who is also the Chief Technical Officer, is regarded in the company as its innovator. Company N has a president and vice-president of operations responsible for approximately 60 employees, a sales vice-president responsible for 11 employees, a quality vice-president responsible for four employees, an engineering vice-president with 10 employees, a CAD director with eight employees and an R&D director with six employees. The company feels that it lives on innovation, both because of its size and because of the very competitive market in which it operates.

Company N's innovation is customer directed; there are specific targets at which it aims innovation and design. Time is limited, and it is also necessary to produce at the right cost. The new product which it is about to launch is a new architecture in silicon technology, one which has been talked about a great deal but has never been executed. It will produce a device one-third to a one-half smaller than former ones, and will lead to significant price reduction. The new architecture is specifically designed around the new CAD tools which are available.

If Company N is to proceed with an innovation, the innovation must, to some extent, conform with the specifications outlined. The company is not interested in making "me too" products, as it does not have the economies of scale to be competitive in this area, and it would have great difficulties in selling such products. The company has to look to the future and to forecast where silicon technology is going. Company N's innovations, it was suggested, must be in semi-customised integrated circuits and must utilise the design tools which the company was specifically set up to take advantage of. This keeps costs down.

Being a small company it thrives on innovative genius. Company N accepts no watering down in the innovation process such as was thought likely to occur in a larger company. All the vice-presidents and directors of departments in the company must be in agreement for an innovation to proceed. Anyone foreseeing a problem must suggest ways in which it may be overcome. The next step in Company N is the planning stage in which objectives are set and agreed; these objectives are expressed in terms of cost, timing, manufacturing and so on. Company N also sets benchmarks for progress. The decision to produce new products in Company N takes not more than 60 days. To get a design down on paper, that is, from the creative idea to the solid concept, the timescale is three to four months, and it will be a year in total before a product is ready to sell.

Another important area in the innovation process in Company N is design creativity. It takes three or four months to transfer a concept into hardware and then it will take three more to have parts produced. This process, it was felt, would take longer in a large company. Both the respondent and the president of Company N had come from a large electronic company.

Financial resources are a very important factor as the company is growing at such a fast rate. Originally it did not have enough capital behind it, and so had to sell equity to raise capital and cash flow problems became evident.

The innovations in the company were dependent on three technologies, the first being CMOS, which has been available for some time but the full benefits have not yet been realised. Company N is capitalising on silicon technology that has been evolving steadily. It feels that the ideal is to pick a point on the edge of manufacturing ability, that is neither too old nor too new and to concentrate work there. A great deal of experience is required to get this just right. Second, CAD (Computer Aided Design) is also a necessary technology for the innovation in Company N. Again Company N works at the technical edge and aims to use CMOS and CAD to a similar level of sophistication so that they are exploited equally and will combine well. The third necessary technology is in computer systems. Here Company N needs systems capable of synchronising methods.

The company is always short of technical personnel and looks for people with professional experience and a good track record. Company N feels that, particularly in a company of its size, it is critical to have a good management team. Typically, the company's vice-presidents have had 15 years experience in the industry at no more than two companies, and have managed at least a department if not a division in their previous company. The management team also has a broad range of skills, and must work in at least two business areas, such as the technical area and the marketing area, or the engineering area and the financial area. Many have degrees in pure science and have gone on to obtain higher degrees in business subjects, such as economics or marketing. At the engineering level, Company N looks for experience, in this case of three to five years, and again a good track record. The company cannot afford to train graduates straight from college, and uses an obvious recruiting hook, the offer of stock options. Most of the senior management in Company N have had to take cuts in their remuneration, often very large cuts, so some other inducement to tempt them was thought necessary. The type of people wanted in the company are difficult to find, although recruiting personnel at the engineering level was thought not quite so difficult. In the early days Company N needed design talent, financial acumen and the marketing skills of experienced people.

The supply of key items is not usually a problem in innovation, even though the company's demands are growing quickly. The company contracts tasks out rather than doing them all in-house. At one time it contracted out testing and fabricating but, as the company has grown, these activities have now been brought in-house and Company N can now finish wafers. This progress continues as the company grows.

Company N regards its relationships with the outside world as to some extent fixed since it is capable of changing only a limited amount of this inter-action. For the Company N respondent, what goes on inside the company in innovation can be very flexible and is strictly creative. This they relate to the requirement for 15 years experience in a senior management team. It was felt that this amount of experience allowed people to know both how to relate to the outside world and also how to change things inside the company.

The marketing assessments which the company makes for innovation are extremely detailed and concern the estimated percentage growth in markets in dollars for 5 to 10 year periods. This process is done for several customer areas, for example, telecoms, and consumers; then the largest and most diverse of these customer areas is broken down further. Company N ranks markets and looks for those with greatest growth potential and aims its product there. There is less interest in possible customers below specified growth thresholds. Company N also looks as the technologies it uses in relation to the market, to see whether there are likely to be changes over a period of time. Ultimately it was felt that it is not possible for a company to do everything, and although marketing must affect decisions, it is necessary to use judgement on where to stop. Achieving perfection is not considered possible, nor is it considered commercially necessary. The aim is to achieve 98% in a year, and this makes the decisions on every project difficult.

Cost assessments in Company N are similarly detailed and are made for each product line. Company N analyses the market to see at what price it thinks it can sell its product and to find out what quantities it can expect to sell. This average selling price is calculated for different market segments. The average selling price and the project costs are analysed to see whether costs are too high. A major uncertainty in Company N is the number of dies on each wafer, as everything else is well controlled. The company recognises the ever present risk that this assessment process will go very wrong. A mis-calculation, of say, the size of a possible market would alter the accuracy of all the other calculations. Also crucial to the costing is the accuracy of the assessment of the time needed to develop the project. Clearly if the project runs behind time, it will cost a great deal more to get into the market than anticipated.

Company N, unlike Company M, feels that secrecy is important to the development of an innovation. It sees itself as rushing for copyright on devices, since once they are produced, they are very easy to copy. To stay ahead, it is necessary that they should be continually upgraded.

Once an innovation decision has been made by the management group, objectives are set and there are monthly and weekly reviews on projects. The company is careful to look out for possible blocks and stoppers and to see whether it needs to dedicate new resouces to the project. Once the project has been defined, a real whole-hearted commitment to it is regarded as essential and cash flow and capital investment must be carefully managed. The company also regards as important not changing the objectives of a project, unless there is an extremely good reason. The Company N respondent saw a distinction between an innovator and somebody with good management skills, but he found it difficult to pinpoint what is necessary to be an innovator. It was felt unproductive to analyse innovators' backgrounds in an attempt to discover the necessary element in being innovative.

Original innovative ideas in Company N come mostly from individuals who have a technical background. Innovators were seen to work to a different timescale from the rest of the people in the company: "They work 20 hours at a time and then you don't see them for 2 days". Once a Company N innovator has had an idea, he must be capable of working with a team, and must be able to use people around him to complete the concept and get it into the hardware stage. The company aims to surround the innovator with those who can pick up his or her ideas and execute them. Innovation champions, it was suggested, should fulfill this sort of role and hence can be very important. People who enjoy this end of the business and who conceptualise on a different level are vital, but the position of the ideal innovation champion in the company depends on the nature of the company. If it is a financially dominated company, an innovation champion deep in the R&D function will have little ability to overcome the organisational constraints which are likely to kill ideas. Above all, innovators must be allowed intellectual freedom.

In Company N governemnt intervention was seen rather less positively than in Company M. Such intervention, it was felt, could be negatively related to innovation, because if government policy changed, whole projects could be destroyed. Company N requires vast sums of money to fuel its fast growth rate (at the moment 50% to 80% a year) and this remains a perennial problem.

For the Company N respondent innovation requires hard work, and the company philosophy is that one is more likely to get this drive from people if they are working for their own company and generating their own rewards, so the ability to give stock options to employees was stressed as important.

For the Company N respondent a commitment to innovation must be clearly understood throughout the organisation, and everyone must know the company objectives. Knowing when to stop can be equally important. It is possible to innovate in many areas and it is difficult but necessary to limit these to the range that the company has decided upon. For Company N an innovative organisation must provide an atmosphere in which there is give and take. No one person can be allowed to dominate. Everyone in Company N has a good track record and this makes them more obviously equal in innovation discussions. The general company environment, or corporate culture, must allow people to think and work in an innovative way and this attitude must come from the top. People must be encouraged to be self-motivated, self-confident enough to ride storms, and not likely to over-react to setbacks. Not everybody needs to agree with an idea at the early stages to ensure its success. It is thought healthy in Company N to have heated discussions. What was important was the general consensus reached at the end of these preliminary discussions. In Company N's philosophy it was also seen as important that when something needs to be done, whoever is available must do it. An example was given of two vice-presidents and another company employee spending one night typing because they had to give a presentation the next day.

APPENDIX 1

List of Participating Companies

Beechams Pharmaceuticals
Bristol-Myers Co.
California Devices Inc.
Ciba-Geigy (UK)
Cincinnati Milacron
Commercial Union Assurance
Consolidated Bathurst
Crownlife Insurance Company
Exxon
Diasonics Inc.
Ferranti
Friends Provident Life Office
GEC
Glaxo
ICI
Inco Ltd.
Ingersoll Rand
London Life
Micro Circuit Engineering
Molins PLC
New England Mutual Life
Northern Telecom Ltd.
Oxford Instruments Group
Plessey Co.
Racal Electronics
RCA Corp.
Reed International
RTZ
Shell UK Ltd.
Sterling Drug Inc.
Thorn EMI
Unilever Limited
Union Oil Company of California
Wellcome Foundation
Westinghouse Electronic Corp.
Wiggins Teape Group Ltd.
L. Williams Teweles & Co.

APPENDIX II

The British-North American Committee Project on Innovation

QUESTIONNAIRE

1 Could you give me a general description of innovation in your company and of the last significant innovation you have introduced?

2 What factors affect decisions to innovate?
—to proceed?
—to defer?
—not to proceed with the development of the innovation?

3 How long do decisions about the innovation process take?

4 What factors affect the speed of progress once an innovation has been adopted?

5 What technologies are essential to your last innovation, and where do these technologies come from?

6 Are there any problems with the adequacy or speed of supply of key items or personnel?

7 What degree of latitude is there in the design and specification of innovations? How are decisions on this made?

8 What market assessment does the company make? And what flexibility is built into the market plan?

9 What cost assessment does the company make? What are the major points of uncertainty about costs?

10 Is secrecy important to success in developing a new innovation such as you have described?

11 What is the internal decision-making process involved in the development of an innovation?

12 What is the career background and training of the senior personnel involved in innovation decisions?

13 Do you think a major champion for an innovation is important?

14 Apart from the "general economic climate", what factors would you rate as highly relevant in the innovation process?

15 What internal factors are significant in being an innovative company?

Suggestions for question 14

—General economic climate (what factors, in particular?)
—General financial position of company
—Specific problems of finance for this scheme
—Actions of competitors
—Other market changes affecting the product/process (direct)

. . . *continued overleaf*

—Supply of qualified personnel:
  1 scientists
  2 technologists
  3 technicians
  4 other personnel
—Supply of key materials or components
—Shop-floor problems or the need to reorganise
—Management problems or the need to reorganise
—Environmental regulations
—Other government intervention
—Unexpected technical problems
—Unexpected cost changes
—Competition within the firm from other projects
—The influence of suppliers and customers
—Presence or absence of protection
—Need to fight new competition
—Any other factors

APPENDIX III

R&D Questionnaire Used in Company Visits

1 What is R&D? Definitions used and differences between R&D.

2 How is R&D within the company organised and structured? What is the employment profile? (Please give numbers and qualifications of R&D personnel.)

3 Why does your company do in-house R&D?

4 Has the balance between in-house and contract R&D changed in the last decade? if so, why?

5 How do you justify returns from R&D investment?
e.g. —scientific return;
—prestige;
—attracting good brains;
—momentum of the past.

6 With what is your major R&D effort concerned with now? Over the past decade?

7 Who decides the direction of the R&D effort?

8 How much are you spending, and how has this changed in the last decade? (Please give the absolute amount and the balance of interests.)

9 How is this amount arrived at and by whom?

10 How has the recession affected this? If there have been cuts, are they more or less than the decrease in turnover? Are there all-round cuts or do they result from a policy decision on a cost/benefit analysis?

11 Do such economies include:
—shortened time horizons?
—reduced range of R&D?
—altered mix of contract/in-house effort?

12 Have there been any other major influences in this period?

13 If you had no pre-existing R&D function, what ideally would you have?

14 How would you describe and rate the primary and secondary functions of the R&D effort in your company?

15 How do the R&D function and its personnel relate to other company functions?

16 How does your R&D activity compare with that of your competitors, nationally and internationally?

17 Do you read any literature in the innovation area, and if so, does it have any policy influence?

Members of the British-North American Committee

Chairmen

A. B. MARSHALL
Chairman, Commercial Union Assurance, London

WILLIAM L. WEARLY
Chairman (Retired), Ingersoll-Rand Company, Woodcliff Lake, New Jersey

Chairman, Executive Committee

WILLIAM I. M. TURNER, JR.
Chairman and Chief Executive Officer, Consolidated-Bathurst Inc., Montreal, Quebec

Members

ROBIN W. ADAM
Chairman, MEPC, London

J. A. ARMSTRONG
Toronto, Ontario

DR. DAVID ATTERTON
Chairman, Foseco Minsep Ltd., Birmingham

CHARLES F. BAIRD
Chairman and Chief Executive Officer, INCO Limited, Toronto, Ontario

JOSEPH E. BAIRD
Los Angeles, California

PROFESSOR SIR JAMES BALL
Chairman, Legal & General Assurance Group plc., London

ROBERT A. BANDEEN
Chairman and President, Crown Life Insurance Company, Toronto, Ontario

*NICHOLAS H. BARING
Managing Director, Baring Brothers Co., Ltd., London

SIR DONALD BARRON
Chairman, Midland Bank plc., London

LAURENT BEAUDOIN
Chairman and Chief Executive Officer, Bombardier Inc. Montreal, Quebec

MICHEL BELANGER
Chairman of the Board, President and Chief Executive Officer, National Bank of Canada, Montreal, Quebec

W. J. BENSON
Deputy Chairman, National Westminster Bank plc., London

C. FRED BERGSTEN
Director, Institute for International Economics, Washington, DC

ROGER BEXON
Deputy Chairman, The British Petroleum Company, plc., London

CARROL D. BOLEN
Vice-President, Pioneer Hi-Bred International Inc., Des Moines, Iowa

THORNTON F. BRADSHAW
Chairman of the Board and Chief Executive Officer, RCA Corporation, New York, NY

F. S. BURBIDGE
Chairman and Chief Executive Officer, Canadian-Pacific Limited, Montreal, Quebec

SIR RICHARD BUTLER
President, National Farmers' Union, London

VISCOUNT CALDECOTE
Chairman, Investors in Industry Group plc., London

SILAS S. CATHCART
Chairman, Illinois Tool Works Inc., Chicago, Illinois

GEORGE J. CLARK
Executive Vice President, Citibank, N.A., New York, NY

FRANK J. CONNOR
President, American Can Company, Greenwich, Connecticut

DONALD M. COX
Director and Senior Vice President, Exxon Corporation, New York, NY

JAMES E. CUNNINGHAM
Chairman and Chief Executive Officer, McDermott International, Inc., New Orleans, Louisiana

JAMES W. DAVANT
Retired Chairman, Paine Webber Incorporated, New York, NY

RALPH P. DAVIDSON
Chairman, Time Incorporated, New York, NY

WILLIAM DODGE
Ottawa, Ontario

**Became a member of the Committee after the Statement was circulated for signature.*

WILLIAM H. DONALDSON
Chairman and Chief Executive, Donaldson Enterprises Inc., New York, NY

EDWARD DONLEY
Chairman, Air Products and Chemicals, Inc., Allentown, Pennsylvania

ALLAN R. DRAGONE
Vice-Chairman, Burlington Industries, Inc., New York, NY

JOHN DU CANE
Director, AMAX Inc., London

*SIR WILLIAM DUNCAN
Chairman, Rolls Royce Limited, London

KEN DURHAM
Chairman, Unilever plc., London

GERRY EASTWOOD
General Secretary, Association of Patternmakers and Allied Craftsmen, London

HARRY E. EKBLOM
Chairman, University Science Partners Inc., Ridgewood, NJ

GLENN FLATEN
President, Canadian Federation of Agriculture, Regina, Saskatchewan

C. S. FLENNIKEN
Chairman, President and Chief Executive Officer, CIP Inc., Montreal, Quebec

DAVID W. FOX
Vice Chairman, The Northern Trust Co., Chicago

SIR ALISTAIR FRAME
Deputy Chairman and Chief Executive, Rio-Tinto Zinc Corporation, London

SIR CAMPBELL FRASER
President, Dunlop Holdings plc., London

ROBERT R. FREDERICK
President, RCA Corporation, New York, NY

GWAIN H. GILLESPIE
Executive Vice-President, Finance and Administration, Chief Financial Officer, R. J. Reynolds Industries, Inc., Winston Salem, NC

MALCOLM GLENN
Executive Vice-President, Reed Holdings Incorporated, Rickmansworth, Herts

PETER A. GRAHAM
Senior Deputy Chairman, Standard Chartered Bank plc., London

JOHN H. HALE
Managing Director, S. Pearson & Son plc., London

JOHN C. HALEY
Executive Vice-President, The Chase Manhatten Bank, New York, NY

HON. HENRY HANKEY
British Secretary, BNAC, Westerham, Kent

AUGUSTIN S. HART, JR.
Retired Vice-Chairman, Quaker Oats Company, Chicago, Illinois

FRED L. HARTLEY
Chairman and President, Union Oil Company of California, Los Angeles, California

H. J. HAYNES
Senior Counselor, Bechtel Group, Inc., San Francisco, California

PAUL HAZEN
Vice Chairman, Wells Fargo & Company, San Francisco, California

G. R. HEFFERNAN
President, Co-Steel International Ltd., Whitby, Ontario

ROBERT HENDERSON
Chairman, Kleinwort Benson, Lonsdale plc., London

*PAUL J. HOENMANS
President, Marketing & Refining Division, Mobil Oil Corp., New York, NY

SIR TREVOR HOLDSWORTH
Chairman, Guest Keen & Nettlefolds plc., London

HENDRIK S. HOUTHAKKER
Professor of Economics, Harvard University, Cambridge, Massachusetts

DONALD P. JACOBS
Dean, J. L. Kellogg Graduate School of Management, Northwestern University, Evanston, Illinois

GEORGE S. JOHNSTON
President, Scudder, Stevens & Clark, New York, NY

PAUL J. KOFMEHL
Director General, Operations, IBM Europe, Paris

**Became a member of the Committee after the Statement was circulated for signature.*

C. CALVERT KNUDSEN
Vice Chairman and Chairman, Executive Committee, MacMillan, Bloedel Limited, Vancouver, B.C.

GAVIN LAIRD
General Secretary, Amalgamated Union of Engineering Workers (Engineering Section), London

PETER LAISTER
Chairman and Chief Executive, Thorn EMI, London

H. U. A. LAMBERT
Chairman, Barclays Bank UK Ltd., London

DR. INGRAM LENTON
Managing Director, Bowater Corporation plc., London

WILLIAM A. LIFFERS
Vice Chairman, American Cyanamid Company, Wayne, New Jersey

WALTER F. LIGHT
Chairman and Chief Executive Officer, Northern Telecom Ltd., Mississauga, Ontario

FRANKLIN A. LINDSAY
Chairman, Engenics Corporation, Lincoln, Massachusetts

SIR PETER MACADAM
Former Chairman, B.A.T. Industries plc., London

IAN MacGREGOR
Honorary Chairman, AMAX Inc., Greenwich, Connecticut

CARGILL MacMILLAN, JR.
Senior Vice President, Cargill Inc., Minneapolis, Minnesota

*JOHN N. MALTBY
Chairman, Burmah Oil plc., Swindon

J. P. MANN
Deputy Chairman, United Biscuits (Holdings) Ltd., Middlesex

PAUL E. MARTIN
President and Chief Executive Officer, The CSL Group, Inc., Montreal, Quebec

WILLIAM J. McDONOUGH
Executive Vice President and Chief Financial Officer, The First National Bank of Chicago, Chicago, Illinois

DONALD K. McIVOR
Chairman and Chief Executive Officer, Imperial Oil Limited, Toronto, Ontario

DONALD E. MEADS
Chairman and President, Carver Associates, Plymouth Meeting, Pennsylvania

SIR PATRICK MEANEY
Chairman, The Rank Organisation plc., London

JOHN MILLER
Vice Chairman, National Planning Association, Washington, DC

WILLIAM R. MILLER
President, Pharmaceutical & Nutritional Group, Bristol-Myers Company, New York, NY

DAVID MORTON
President and Chief Executive Officer, Aluminium Co. of Canada Ltd., Montreal, Quebec

JOHN J. MURPHY
Chairman, President & Chief Executive Officer, Dresser Industries, Dallas, Texas

KENNETH D. NADEN
President, Volunteers in Overseas Cooperative Assistance, Washington, DC

TED NEWALL
Chairman, President and Chief Executive Officer, Du Pont Canada Inc., Mississauga, Ontario

WILLIAM S. OGDEN
Chairman, Continental Illinois National Bank and Trust Company of Chicago, Chicago, Illinois

ANTHONY J. E. O'REILLY
President and Chief Executive Officer, H. J. Heinz Company, Pittsburgh, Pennsylvania

LORD PENNOCK
Chairman, BICC plc., London

*D. A. S. PLASTOW
Managing Director, Vickers plc., London

GEORGE J. POULIN
General Vice President, International Association of Machinists & Aerospace Workers, Washington, DC

SIR RICHARD POWELL
Hill Samuel Group Limited, London

ALFRED POWIS
Chairman and Chief Executive Officer, Noranda Mines Limited, Toronto, Ontario

---

**Became a member of the Committee after the Statement was circulated for signature.*

PAUL E. PRICE
Executive Vice-President, International Grocery Products, The Quaker Oats Company, Chicago, Illinois

GRANT L. REUBER
President, Bank of Montreal, Toronto, Ontario

BEN ROBERTS
Professor of Industrial Relations, London School of Economics, London

HAROLD B. ROSE
Group Economic Adviser, Barclays Bank plc., London

CHESTER A, SADLOW
Executive Vice-President, Advanced Production Technology, Westinghouse Electric Corp., Pittsburgh, Pennsylvania

WILLIAM R. SALOMON
Honorary Chairman, Salomon Brothers, Inc., New York, NY

A. C. I. SAMUEL
Handcross, Sussex

HOWARD D. SAMUEL
President, Industrial Union Department, AFL-CIO, Washington, DC

NATHANIEL SAMUELS
Advisory Director, Lehman Brothers Kuhn Loeb Inc., New York, NY

*THOMAS R. SAYLOR
Vice-President, Garnac Grain Co., Inc., New York, NY

LORD SEEBOLM
Dedham, Essex

THE EARL OF SELKIRK
President, Royal Central Asian Society, London

JACK SHEINKMAN
Secretary-Treasurer, Amalgamated Clothing & Textile Workers' Union, New York, NY

LORD SHERFIELD
Chancellor of Reading University, London

R. MICHAEL SHIELDS
Managing Director, Associated Newspapers Group, plc., London

GEORGE L. SHINN
Chairman and Chief Executive Officer Emeritus, The First Boston Corporation, New York, NY

GORDON R. SIMPSON
Chairman, General Accident Fire and Life Assurance Corporation plc., Perth, Scotland

SIR ROY SISSON
Chairman, Smiths Industries plc., London

SIR LESLIE SMITH
Chairman, BOC International plc., London

RALPH I. STRAUS
New York, NY

THOMAS H. B. SYMONS
Vanier Professor, Trent University, and Director, Celanese Canada, Inc., Peterborough, Ontario

ALLAN R. TAYLOR
President and Chief Operating Officer, The Royal Bank of Canada, Toronto, Ontario

JAMES O. THACKRAY
Chairman, Bell Canada, Montreal, Quebec

W. C. THOMSON
A Managing Director, Royal Dutch/Shell Group of Companies, London

ALEXANDER C. TOMLINSON
President, National Planning Association, Washington, DC

ALAN TUFFIN
General Secretary, Union of Communication Workers, London

J. C. TURNER
General President, International Union of Operating Engineers, AFL-CIO, Washington, DC

W. O. TWAITS
Toronto, Ontario

JOHN A URQUHART
Executive Vice President and Sector Executive, International Sector, General Electric Company, Fairfield, Connecticut

MARTHA REDFIELD WALLACE
President, Redfield Associates, New York, NY

GLENN E. WATTS
President, Communications Workers of America, AFL-CIO, Washington, DC

VISCOUNT WEIR
Chairman, The Weir Group Limited, Glasgow, Scotland

**Became a member of the Committee after the Statement was circulated for signature.*

FREDERICK B. WHITTEMORE
Managing Director, Morgan Stanley & Co. Incorporated, New York, NY

MARGARET S. WILSON
Chairman of the Board, Scarbroughs, Austin, Texas

CHARLES WOOTTON
Director, International Public Affairs, Gulf Oil Corporation, Pittsburgh, Pennsylvania

Sponsoring Organizations

The British-North American Research Association was inaugurated in December 1969. Its primary purpose is to sponsor research on British-North American economic relations in association with the British-North American Committee. Publications of the British-North American Research Association as well as publications of the British-North American Committee are available from the Association's office, 1 Gough Square, London EC4A 3DE (Tel. 01-353 6371). The Association is recognized as a charity and is governed by a Council under the chairmanship of A. B. Marshall.

The National Planning Association is an independent, private, nonprofit, nonpolitical organization that carries on research and policy formulation in the public interest. NPA was founded during the Great Depression of the 1930s when conflicts among the major economic groups—business, labor, agriculture—threatened to paralyze national decision making on the critical issues confronting American society. It was dedicated to the task of getting these diverse groups to work together to narrow areas of controversy and broaden areas of agreement and to provide on specific problems concrete programs for action planned in the best traditions of a functioning democracy. Such democratic planning, NPA believes, involves the development of effective governmental and private policies and programs not only by official agencies but also through the independent initiative and cooperation of the main private-sector groups concerned. And to preserve and strengthen American political and economic democracy, the necessary government actions have to be consistent with, and stimulate the support of, a dynamic private sector.

NPA brings together influential and knowledgeable leaders from business, labor, agriculture, and the applied and academic professions to serve on policy committees. These committees identify emerging problems confronting the nation at home and abroad and seek to develop and agree upon policies and programs for coping with them. The research and writing for these committees are provided by NPA's professional staff and, as required, by outside experts.

In addition, NPA's professional staff undertakes research designed to provide data and ideas for policymakers and planners in government and the private sector. These activities include the preparation on a regular basis of economic and demographic projections for the national economy, regions, states, metropolitan areas, and counties; research on national goals and priorities, productivity and economic growth, welfare and dependency problems, employment and manpower needs, and energy and environmental questions; analyses and forecasts of changing international realities and their implications for U.S. policies; and analyses of important new economic, social and political realities confronting American society.

NPA publications, including those of the British-North American Committee, can be obtained from the Association's office, 1606 New Hampshire Avenue, N.W., Washington, D.C. 20009 (Tel. 202-265-7685).

The C. D. Howe Institute was established in 1973 by the merger of the C. D. Howe Memorial Foundation and the Private Planning Association of Canada. It is a nonprofit, nonpolitical organization seeking to contribute nonpartisan research findings and commentary on Canadian economic policy issues.

The guiding principle of the Institute is to conduct its research and analysis in a manner that is balanced in approach, professional in method and readable in style.

To ensure diversity in perspective, participation is encouraged from organized labor, business, agricultural associations, and the professions.

While its focus is national, the Institute recognizes that Canada is composed of regions, each of which may have a particular point of view on policy issues, unique interests and concerns, and different concepts of national priorities. The Institute also pursues involvement from both of Canada's major linguistic communities.

It is not the purpose of the Institute to promote consensus on policy issues, although on occasion that may be feasible. The primary function is to add to public understanding of issues by providing sound analysis reflecting objective treatment of diverse points of view. The staff of the Institute seeks to develop good working relationships with public officials for the purpose of better understanding the basis for government decisions and contributing effectively to public policy formulation.

A Board of Directors is responsible for the general administration of the Institute, while the immediate direction of the policies, program and staff is vested in the Executive Director, The function of the Board is to make independent research and publication possible under the most favorable conditions and not to control the conduct and conclusions of research activity.

Mr. Michael Belanger is Chairman, Dr. Wendy Dobson is Executive Director and Treasurer, and Ms. Wendy Walker is Corporate Secretary.

The Institute's publications, including those of the British-North American Committee, are available from its offices: Suite 900, 555 Dorchester Blvd., Montreal, Quebec H2Z 1B1; P.O. Box 1621, Calgary, Alberta T2P 2L7; and Glendon Hall, 2275 Bayview Avenue, Toronto, Ontario M4N 3M6 (Tel. 416-485-4310).

Publications of the British-North American Committee

BN-34 *The Process of Innovation,* by Nuala Swords-Isherwood, October 1984 (£5.00, US$8.00, CAN$10.00)

BN-33 *Governments and Multinationals: Policies in the Developed Countries,* by A. E. Safarian, December 1983 (£4.00, $8.00)

BN-32 *Trade Issues in the Mid 1980s,* by Sidney Golt and a Committee Policy Statement, October 1982 (£3.50, $7.00)

BN-31 *The Newly Industrializing Countries: Adjusting to Success,* by Neil McMullen, November 1982 (£3.50, $7.00)

BN-30 *Conflicts of National Laws with International Business Activity: Issues of Extraterritoriality,* by A. H. Hermann, August 1982 (£3.00, $6.00)

BN-29 *Industrial Innovation in the United Kingdom, Canada and the United States,* by Kerry Schott, July 1981 (£2.25, $5.00)

BN-28 *Flexible Exchange Rates and International Business,* by John M. Blin, Stuart I. Greenbaum and Donald P. Jacobs, December 1981 (£3.00, $8.00)

BN-27 *A Trade Union View of US Manpower Policy,* by William W. Winpisinger, April 1980 (£1.75, $3.00)

BN-26 *A Positive Approach to the International Economic Order, Part II: Non-trade Issues,* by Alasdair MacBean and V. N. Balasubramanyam, May 1980 (£2.25, $5.00)

BN-25 *New Patterns of World Mineral Development,* by Raymond F. Mikesell, September 1979 (£2.25, $5.00)

BN-24 *Inflation is a Social Malady,* by Carl Beigie, March 1979 (£2.00, $4.00)

BN-23 *A Positive Approach to the International Economic Order, Part I: Trade & Structural Adjustment,* by Alasdair MacBean, October 1978 (£1.75, $3.00)

BN-22 *The GATT Negotiations 1973-79: The Closing Stage,* by Sidney Golt and a Committee Policy Statement, May 1978 (£1.50, $3.00)

BN-21 *Skilled Labour Supply Imbalances: The Canadian Experience,* by William Dodge, November 1977 (£1.50, $3.00)

BN-20 *The Soviet Impact on World Grain Trade,* by D. Gale Johnson, May 1977 (£1.75, $3.00)

BN-19 *Mineral Development in the Eighties: Prospects and Problems,* a Report Prepared by a Group of Committee Members with a Statistical Annex by Sperry Lea, November 1976 (£1.50, $3.00)

BN-18 *Skilled Labour Shortages in the United Kingdom: With Particular Reference to the Engineering Industry,* by Gerry Eastwood, October 1976 (£1.50, $3.00)

BN-17 *Higher Oil Prices: Worldwide Financial Implications,* a Policy Statement by the British-North American Committee and a Research Report by Sperry Lea, October 1975 (£1.50, $3.00)

BN-16 *Completing the GATT: Toward New International Rules to Govern Export Controls,* by Fred Bergsten, October 1974 (80p, $2.00)

BN-15 *Foreign Direct Investment in the United States: Opportunities and Impediments,* by Simon Webley, September 1974 (80p, $2.00)

BN-14 *The GATT Negotiations, 1973-75: A Guide to the Issues,* by Sidney Golt, April 1974 (£1.00, $2.00)

BN-13 *Problems of Economic Development in the Caribbean,* by David Powell, compiled from a study by Irene Hawkins, November 1973 (80p, $2.00)

BN-12 *The European Approach to Worker-Management Relationships,* by Innis Macbeath, October 1973 (£1.00, $2.50)

BN-11 *An International Grain Reserve Policy,* by Timothy Josling, July 1973 (40p, $1.00)

BN-10 *Man and His Environment,* by Harry G. Johnson, August 1973 (40p, $1.00)

BN-9 *Prospective Changes in the World Trade and Monetary System: A Comment,* A Statement by the BNAC, October 1972 (30p, $.75)

BN-8 *Multinational Corporations in Developed Countries: A Review of Recent Research and Policy Thinking,* by Sperry Lea and Simon Webley, March 1973 (80p, $2.00)

BN-7 *Sterling, European Monetary Unification, and the International Monetary System,* by Richard N. Cooper, March 1972 (40p, $1.00)

BN-6 *Multinational Corporations and British Labour: A Review of Attitudes and Responses,* by John Gennard, January 1972 (80p, $2.00)

BN-5 *The Strategic and Political Issues Facing America, Britain and Canada,* by Leonard Beaton, October 1971 (40p, $1.00)

BN-4 *Purposes and Prospects,* A Policy Statement describing the BNAC, April 1971 (20p, $.50)

BN-3 *British Entry to the European Community: Implications for British and North American Agriculture,* by John S. Marsh, together with *Agriculture Policies for World Trade Expansion,* A Statement by the BNAC, March 1971 (50p, $1.25)

BN-2 *Transatlantic Relations in the Prospect of an Enlarged European Community,* by Theodore Geiger, November 1970 (60p, $1.50)

BN-1 *An Overall View of International Economic Questions Facing Britain, the United States and Canada during the 1970s,* by Harry G. Johnson, June 1970 (40p, $1.00)

Occasional Paper-3 *Managing Product Innovation,* The views of five BNAC members: Carrol Bolen, Ken Durham, Walter Light, Chester Sadlow and Viscount Weir, October 1984 (£1.50 US$2.00, CAN$2.50)

Occasional Paper-2 *Managing the Response to Industrial Decline,* by Joseph L. Bower, September 1984 (£1.50, US$2.00, CAN$2.50)

Occasional Paper-1 *New Investment in Basic Industries,* prepared by a Committee Task Force, June 1979 (60p, $1.00)

Publications of the British-North American Committee are available from:

In Great Britain and Europe
BRITISH-NORTH AMERICAN RESEARCH ASSOCIATION
Grosvenor Gardens House, 35-37 Grosvenor Gardens, London SW1W 0BS
Telephone: 01-828 6644

In the United States of America
NATIONAL PLANNING ASSOCIATION
1606 New Hampshire Avenue, NW
Washington, DC 20009
Telephone: 202-265-7685

In Canada
C. D. HOWE INSTITUTE
2275 Bayview Avenue, Toronto, Ontario M4N 3M6
Telephone: 416-485-4310